Science and Culture in Europe

British Library Cataloguing-in-Publication Data
A catalogue record for this publication is available from the British Library

Cover artwork and illustrations by David Newton, artist-in-residence at the
Science Museum, London, Summer 1993.
Cover design by the Brickworks Design Consultants.
Translations into English by the Department of English, University of Nice
(Margaret Campbell, Kevin Lemoine, Ruth Pinnegar and Alison Wilshaw)
and First Edition Translations Ltd.
Set from Pagemaker in Postscript Monotype Times New Roman.
Printed in England by Antony Rowe, Chippenham, Wilts.

Articles © their authors 1993
Compilation © Trustees of the Science Museum 1993

ISBN 0 901805 66 1

For further information about journal subscriptions, please contact:
Alliage, 78 route de Saint-Pierre de Féric, F-06000 Nice, France
Public Understanding of Science, Science Museum Library, London SW7 5NH

Science Museum, Exhibition Road, London SW7 2DD

Science and Culture in Europe

conceived and commissioned by
Jean-Marc Lévy-Leblond

English-language edition edited by
John Durant and Jane Gregory

with forewords by
Rt Hon William Waldegrave
Antonio Ruberti
Sir Mark Richmond
Hubert Curien

Science Museum
on behalf of *Alliage* and *Public Understanding of Science*

Acknowledgements

We are grateful to the Office of Science and Technology and the Science and Engineering Research Council for supporting the production of this English-language edition of *Science & Culture en Europe*, a special issue of the French journal of science and culture *Alliage*. The French edition was supported by DGXII of the Commission of the European Communities and the Cultural Affairs Office for the Provence, Alps and Côte d'Azur region. The project was conceived by Jean-Marc Lévy-Leblond, and we are grateful to him, and to the staff and sponsors of *Alliage*, for enabling us to bring it to an English-language readership.

John Durant
Editor, *Public Understanding of Science*

Jane Gregory
Managing Editor, *Public Understanding of Science*

Contents

The challenge

Jean-Marc Lévy-Leblond
Editor, *Alliage*

It is in Europe that the mixture of barbarism, technical expertise and science has been concocted which is now unfurling in our global iron age (...). Totalitarianism is a European invention three times over. Nuclear extermination, although American by birth, is of European genealogy. And it was that pacifist genius, Albert Einstein, who urged the President of the United States of America to manufacture the first atomic bomb. Thus Europe's most original creations have today been universalized, for better and for worse. Reason has expanded in the form of critical rationality, in the form of a rationality based on myth and in the form of an instrumental rationality which serves barbarism. Humanism has spread to allow human rights to be established in numerous places, however at the same time it has allowed the oppression of human beings all over again.[1]

Europe was invented by culture and invented science; the time has come for it to confront these two sides of its identity. Beyond the uncertainties which Europe faces in its quest for economic and political unity, this confrontation is in response to an urgent need. How else will we meet the challenge of modernity – a modernity which is throwing technocracy, the offspring of this science, at democracy, the off-spring of this culture? Will we be able to invent in time new means of future collective management which will prevent public consciousness from being taken over by technical expertise?

From the start, this century has reminded us that cultures are mortal. Before its end, it may well teach us that science is not immortal. Just as the political tensions and economic forces of the not-so-distant past were able to work for the Renaissance of culture as well as the birth of science in Europe, so together they could lead to their decline in an even closer future. The 'defence and illustration' of science and culture, which from now on will of necessity be a shared responsibility, can only be accomplished by Europeans acting together.

Still, a clear vision must be maintained with regard to the complex relations between science and culture on the European scale. It is the current situation, places and activities of Europe – strengths and weaknesses, unity and diversity, developments and projects – that these essays present. This collection, the original version of which is a special issue of the French journal of science and culture *Alliage*, is in itself an example of the new initiatives which from now on will be essential: this English-language edition is published under the auspices of *Public Understanding of Science*, and extracts will appear in Spanish in *Arbor*, in Italian in *Prometeo*, in Dutch in *Iota*, in Portuguese in *Colloquio Ciencia* and in Swedish in *Vest*.

Reference

1   Morin, E., 1990, *Penser l'Europe* (Paris: Gallimard).

Prologue

Science, culture and government

Rt Hon William Waldegrave MP
Minister for Public Service and Science, UK

I am very glad to have the chance to contribute a foreword to this timely volume of essays. But I must begin by picking the editors up on one point: the title. 'Science and Culture in Europe'? I am sure this is a shorthand formula, rather than a statement implying that the two are separable. Nevertheless, it could be read as illustrating a perception that is all too common in the United Kingdom: that culture means the arts, and that the sciences are some mysterious, inferior, and separate activity. Even the BBC, with its world-renowned range of programmes in science and natural history, sometimes appears guilty of this misleading assumption. Science is a mainstream element in European culture. This is I think a point more widely accepted and understood in France than in the United Kingdom. Nevertheless, all of us throughout Europe must work to get this basic truth across.

I am not underestimating the difficulties of the task we are setting ourselves. Take the standard Hollywood portrayal of a scientist: a bizarrely coiffured maniac in a white coat, lurking dangerously in a laboratory full of sinister gadgets and test tubes giving off dense vapour. Add to this the notion of scientists as developed by extreme animal rights groups: sadistic men inflicting immense and avoidable pain on conscious animals. Mix this imagery with the fact that even serious newspapers will not run a non-medical science story unless it has a zany element, and you have a recipe for glib prejudice and ignorance.

We must counteract these perceptions, for practical as well as cultural reasons. Not just so that our citizens understand – as Bruce Durie, Director of the Edinburgh International Science Festival, put it to me – that almost every home in Western Europe has twenty-three microchips, two electric motors, two linear accelerators and a cavity magnetron. The stark truth is that scientists and engineers are at the same time our greatest, and our most undervalued, resource.

There are various, and probably equally valid, ways of addressing this cultural problem. In the United Kingdom we have many good, small-scale projects at the grass roots level, often involving young people. In France, we see a successful higher-profile approach, embodied by the excitement of La Science en Fête. Now we are also witnessing the first stirrings of coordinated action throughout the European community, with a prototype European Week for Scientific Culture in November 1993. Throughout Europe we have the resources – in our museums, in our universities and research organizations, in interactive science centres such as the Eksperimentarium in Copenhagen or Techniquest in Cardiff – to make a real impact.

Is there a role for government in all this? Through the United Kingdom Government's recent White Paper setting out future policy on science and technology, we are aiming to create a framework to bring our scientists centre stage. We wish to develop a systematic partnership between the scientific community, industry and commerce, and Government. This will strengthen the contribution of science and engineering to wealth creation, and quality of life. Implicit in this process, I believe, will be increasing recognition both of the central relevance of science to all our lives, and that scientists are a fundamental asset to any nation.

Reforms of education in the UK, especially developments in the National Curriculum, should also bring real improvement in the level of understanding of scientific issues across the population, and boost the supply of scientifically literate young people entering employment. But these measures will take time to work through, and will not be enough on their own. The UK Government has therefore embarked, together with research charities and other organizations prominent in the field, on a nationwide campaign to promote the public understanding of science. We shall be working across the country on a variety of initiatives to make science more attractive to the nation and the nation's children.

It is on this last resource that all our futures depend. We must work together to convince our children of the beauty, excitement, and fundamental importance of science, and of the central position it occupies, alongside the arts, in the culture of our continent.

Science in European culture

Antonio Ruberti
*Vice-President of the Commission of the European Communities,
responsible for research and education*

Modern science was born in Europe. The concept of a systematic study of nature through the confrontation of hypotheses, observations and controlled experiments has, since the time of Galileo, represented a characteristic element of European culture. Since the industrial revolution, the principle of applying the knowledge collated in this manner to economic development has been a fundamental feature of European society.

In a certain sense science was also born European. Before the springing up of the nation states in the nineteenth century, science developed as a fashion and in a cosmopolitan atmosphere throughout the continent as a result of the movement of individuals, the exchange of scientific correspondence and contact between the leading scholars across national borders.

European in this dual sense, science has long believed in its implicit confidence and capacity to push back the frontiers of knowledge and to continually raise the level of prosperity and well-being.

Since the pioneering era, however, circumstances have changed greatly. Science intervenes in almost every aspect of life in our society and, together with technology, forms a powerful system at the heart of this society which is largely autonomous. Today science and technology have become the concern of very specialized professionals, and are no longer a true part of most people's culture.

Because it gives people the power to act at the heart of matter and life, the advancement of knowledge and technology has also brought a number of ethical and social problems. Moreover, science and technology are required to resolve the problems associated with economic and industrial development in a finite world, and help to sustain a habitable environment on Earth.

On the other hand, over the last few decades Europe has ceased to be the centre for learning and technology. In more than one discipline it is the United States which dominates the world scene, and it is in Japan (and perhaps soon in the greater part of the Pacific) where a large proportion of the advanced technologies have been developed. In other respects also, science is no longer spontaneously European. To a large extent research is financed and organized at a national level and compartmentalization between different national systems has become important.

Faced with this triple evolution, action must be taken. It is imperative that efforts be made to restore the balance of global research to the benefit of Europe, to reconstruct a European scientific community, and to establish new relationships between the scientific and technological system and society which would allow both the resolution of the ethical problems associated with the advancement of knowledge, and the establishment of a lasting model for development.

All this could be achieved simultaneously and in one movement. In order to assist

European research to reassert its importance on the worldwide scientific stage, and similarly, to put society in a better position to resolve environmental and social problems linked to technological and industrial development, all that is needed is to exploit the diversity which is a feature of Europe's richness, against a background of unity. Faced with all the problems mentioned, we must rely on the dual principle of the integration of systems and activities and the optimization of differences.

A scientific and academic European area

First of all this requires the integration of research activities themselves. Thanks to the implementation of a certain number of initiatives, European research is a reality: evidence of this are CERN, the European Centre for Nuclear Research; EMBO, the European Molecular Biology Organization; and ESO, the European Space Organization. Conducted in a structured way for almost ten years, the research programmes of the European Community have contributed to changing profoundly the European scientific and technological landscape. A network of associated laboratories, research centres and companies has been created, a habit of cooperation has been acquired and often permanent links have been established.

Nevertheless one cannot claim that there exists today genuine policy for common research in Europe. Coordination is still weak between the action taken within the framework of national programmes, Community programmes and different European organizations of scientific cooperation. Mechanisms must be developed to allow a genuine 'European scientific and technological area' to be set up which is based upon, but also extends, those efforts already undertaken.

Moreover, recent history has opened up the possibility of extending this area over the entire continent. The disappearance of the East/West divide has enabled Europe to finally reconcile its political and geographical reality. Here is a chance for all of us to grasp that opportunity. With regard to the isolation which scientific communities suffered in Eastern Europe, and the traditions and often original approaches which were developed there and in the republics of the former Soviet Union, we now have the chance to offer support. However, there is a real risk of seeing the existing scientific communities broken up by a spontaneous 'brain drain'. In the long term, it is not only these countries which would suffer from such destruction, but European research as a whole.

Parallel to the establishment of a 'scientific European area', there is a need for the development of an 'academic European area'. Since the Middle Ages universities have represented major centres of learning. The relations between the universities of different European countries no longer resemble those which existed during the medieval period. Today university systems have a strong national character and are organized in very different ways. Basic research remains the privileged domain of universities, and in order to allow research to develop in an optimum way in Europe, integration of the work carried out in European universities in the various disciplines must be reinforced beyond traditional academic links.

Equal emphasis should be placed on the second aspect of the function of universities: teaching. Thanks to the Community programmes ERASMUS and COMETT, today training opportunities in universities in other countries are offered to numerous European students. However, the number of students concerned still remains limited.

But there is in Europe today a genuine pool of titles and qualifications, even though, in terms of academic recognition of training acquired outside the national system, a great deal (if not everything) remains to be done. The aim, of course, is not to standardize training systems – this would be hugely complicated and might even have disastrous consequences. The coexistence of different systems and approaches allows richer and more comprehensive training to be considered. Based on principles of equality and mutual trust which support the conditions governing professional recognition, our aim must be to obtain acceptance of the validity of the final product, the training itself.

At the interface between the problems of research and education, a whole field opens up in terms of the optimization of the often very distinct national experiences and approaches implemented in different countries – that of scientific education and culture. To provide Europe with the new generations of researchers and engineers it will need increasingly in the decades to come, and to guarantee that the scientific and technological choices will be carried out as democratically as possible, it is imperative that efforts be made in this area within Europe, and at the European level.

The aim of the European Week for Scientific Culture which the Commission has initiated and organized for Autumn 1993 is to signal the intensification and harmonization of activities carried out according to this plan in the European countries. An essential aspect of the initiatives which have been set up on this occasion will be the intensification of links between specialist agencies concerned with scientific culture and teaching systems. In the long term, the aim is to offer citizens, decision makers and researchers themselves the opportunity to equip themselves with a genuine scientific and technical culture, and more precisely to restore science and technology to the place they should occupy in culture and which they should never have lost.

Such an approach, which combines the efforts of integration and respects differences, is also essential when faced with the ethical questions linked to scientific and technological development. The perception of ethics varies considerably between countries. Perceptions are deeply rooted in certain cultural and social characteristics which are themselves fashioned by history. For example, how could one understand the attitude of citizens in Germany to biotechnology without an appreciation of the tradition of natural philosophy which characterized German culture in the eighteenth and nineteenth centuries?

In an economically unified Europe where goods and people are able to circulate freely, the coexistence of very distinct regulations is not always possible in practice. In the biomedical sphere, for example, this might culminate in an undesirable migration of researchers, doctors and even dying people. Codes of conduct should therefore be adopted which respect the cultural differences of perception and sensitivity.

The trump card of diversity

Although it is a handicap in some respects, for the most part European diversity represents a trump card. Among the high stakes involved in scientific and technological development in the years to come, the question of North/South relations appears in its different forms: the links between the problems of development and environment; the question of 'appropriate' technologies; and the way in which science and technology can effectively contribute to helping developing countries resolve their serious problems.

For well-known historical reasons, of the three great scientific and technological world powers, Europe is unquestionably the nearest to the countries in the Southern hemisphere. Faced with the characteristics of the societies in developing countries, European cultural diversity represents an additional trump card. The coexistence of different cultures in the same area has in fact – and in spite of how it may seem at times – taught Europeans on the whole to be more open to the values of other societies. Furthermore European development certainly offers a school of tolerance which is particularly effective in this respect.

Certainly Europe is not confining itself to the European Community. However, in terms of European development, the Community represents both the clearest expression and the most powerful instrument required to achieve this. To a large extent, therefore, it is within the Community framework that the problems mentioned can be approached with the best results, and it is often at this level that actions are undertaken with the greatest benefit. The Commission has recently submitted its proposals for a new framework programme of research and technological development for the period 1994–1998 to the Ministers for Research in the Member States and to European parliamentarians. Similarly it has presented its views on the future of the action taken in the area of education and training. For the first time, and on a genuinely significant scale, it is taking initiatives within the sphere of culture and scientific education. A framework has been offered which can be used to establish new, deep-rooted and lasting links between science, culture and society in Europe.

Bridging the divide

Sir Mark Richmond
Chairman, Science and Engineering Research Council, UK

The nineteenth and twentieth centuries have seen prodigious advances in our understanding of the natural world and the principles that guide it, and the application of this knowledge has transformed the living conditions of many on this planet. Nor it is likely that the pace of advance will slow down – indeed, it quickens: already new understanding is foreshadowing new applications, and knowledge builds on knowledge till the march of technology leaves us breathless. But this ever-accelerating progress has been matched by a developing schism. More and more the scientist is divided from his or her fellow human beings; the principles of science are understood by an ever-smaller proportion of our population. There are indeed two cultures, and in much of Europe the divide between them pervades everything: education, administration and institutions.

We will allow this schism to develop further at our peril. Already modern scientific advances raise ethical and humanitarian problems whose handling is made grievously more difficult by the two cultures – by this lack of communication and dialogue. How will we resolve the problems posed by the human genome project? How will we as individuals retain our independence and integrity in a world dominated by faster and faster and larger and larger computers? Where will in vitro fertilization lead us?

For too long the science funders, at least in the UK, have effectively ducked their responsibility to play their part in helping to bridge the two cultures. Too easily it has been assumed that it was the responsibility of non-scientists to make all the effort. Now things are changing – this volume is evidence of that. Yet more must be done, and quickly.

Sharing knowledge

Hubert Curien
President-Elect of CERN and Professor of Physics, Université Paris 6

Some of our contemporaries sometimes boast of their ignorance of science, but they would be annoyed if their words were taken seriously enough to be used as proof of a lack of culture. The erratic snobbishness of self-glorified ignorance does not in itself constitute a real social danger. What is a lot more fearsome is the disabling ignorance that creates an insidious and obvious gap between those who know – or are in the position of acquiring knowledge – and everyone else.

The 'Saint Matthew' effect applies here as it does everywhere else: the learned keep getting more learned (like the rich keep getting richer). The rule is the same as for soap bubbles: two bubbles that stick together do not become the same size, but rather the larger takes over the smaller. The sharing of knowledge cannot be entrusted solely to the forces of nature; it demands a political will. Furthermore, advanced nations have the honour of actively leading the crusade.

Scientific scholarship is a body of knowledge in perpetual evolution. It is also a discipline of the mind, one that logical people call 'structuring' and more cautious people call 'distorting.' If scientific application cannot make claims to universality, what it can do is put human intelligence in a position to create a useful representation of the world. It is permissive. It does not limit the imagination: it provides it with a stepping-stone. Furthermore, if scientists are the hunters of the 'how,' they do not claim to have an explanation for every 'why.' The prime causes of many phenomena still elude us. Some will probably always elude us. Why would our vocation be to understand everything? The essential thing is that we understand as much as possible.

Pic de la Mirandole, a Renaissance European who had studied everything, was, as the saying goes, a paragon of tolerance and openness. He was a good model for learned people from every period of history. Authentic scholarship knows its limits, its weaknesses, and its imperfections. The practice of studying culture is a school of modesty. Mathematicians tell us that for any phenomenon that can be expressed in an equation, it is easier to control it than to predict its behaviour. Controlling means taking pains to continually correct in order to maintain the balance between the model and reality. Predicting means trying to find a long-term, equation-based solution to a problem whose very nature may be to diverge. This is enough to make anyone modest, not only mathematicians.

Roger Caillois, closer to us than Pic de la Mirandole, a philosopher, naturalist, and scholar, classified people's games into four groups: games of chance, games of competition, games of imitation, and games of excitement. Isn't research, which has many of the properties of a game, a collection of all four of these groups? That chance may put the researcher on the right track is beyond doubt, though some people call it intuition. Competition in science is the rule: it serves as an accelerator. Imitation of established masters, indeed of colleagues, is part of current practice and is generally honest. As for the excitement of discovery, no comment is necessary. Anyone who has had even a small taste of it has tasted paradise.

In scientific research everyone can benefit: those who play the games and those who watch. Europeans, play your games, but play them so that everyone may benefit from them.

Overview

Science and culture(s) in Europe

Michel André
DGXII (Science, Research and Development), Commission of the European Communities

What are the relationships between science and culture in Europe? There is no easy answer to this question. To begin with, in which sense are we to take the word 'culture'? In a comment on his famous talk in 1959, 'The Two Cultures', CP Snow said how happy he was to have deliberately used a word which had a double meaning. In fact one can use the word culture in three different ways. In the French, or more generally Latin, sense of the term, culture is the ensemble of an individual's knowledge, including their capacity to judge and to appreciate the products of human creativity. In a more Germanic sense of the term it is also the objective intellectual output of a civilization ('Latin culture', 'Chinese culture', etc.). Finally, in a more 'anthropological' and Anglo-Saxon sense of culture, it is the system of values, attitudes, thought processes, behaviour and so on of a given social group.

Three categories of questions correspond to these three different meanings, which are undoubtedly connected in practice even though they are quite distinct. All three are dealt with in this issue. For example, in the articles by John Durant on the theme of what it means to be scientifically literate, by Franco Prattico on the knowledge of science in Italy, and by Charles Tanford and Jacqueline Reynolds on scientific tourism in Europe, it is culture of the first kind that is being considered. On the other hand the analysis by Paolo Galluzzi of the problems of developing the scientific heritage in Italian museums is to be associated with the second sense of the word culture. Finally it is the anthropological meaning which is at work in Goéry Delacôte's discussion of American culture or when Michiel Schwarz speaks of 'technological culture'.

These three aspects certainly cannot be separated from each other and it is difficult to treat them in isolation. The article by Renate Bader on the situation in Germany is exemplary from this point of view. The place of science in the culture of German citizens (meaning number 1) emerges there tightly linked to the situation in German culture (meaning number 2), which is itself a product and an expression of values, perceptions and attitudes which are typical of Germanic culture as such (meaning number 3). To one extent or another, this close linkage between the three dimensions of culture underpins several contributions to this book.

A complex picture full of contrasts

If the question of the relationships between science and culture in Europe is complex, it is also because of the great variety of situations with which one has to deal here. In this field even more than in many others, Europe is actually far from being a homogeneous whole: in different countries the problems are perceived and the questions posed in very different ways.

The differences in question are firstly differences in the concepts used. The best example is the expression 'scientific and technical culture'. Used widely in France and in the Latin world, it does not have a strict equivalent in other languages. Terms such as 'public understanding of science' or 'scientific literacy' which are used in the Anglo-Saxon world are indeed trying to grasp the same realities. However, since they focus on the problem of the knowledge and understanding of science by the public at large, they do not take into account aspects related to the control by scientists themselves of the intellectual and historical roots of their discipline (do researchers today have a real 'scientific culture' of their own?). The best approximation to the same category of problems in German is *Wissenschaftsbildung*. However, its clear educational connotation gives it a rather special significance.

These linguistic distinctions may seem to be superficial. In fact they are rather basic. Without being real obstacles to mutual understanding, they do help us to grasp the different ways in which the relationships between science and culture are understood in different 'cultures' (sense number 3). As the linguists Wilhelm von Humboldt, Edward Sapir and BL Whorf have shown, a language is not just a simple tool. Each language expresses a view of the world, and in using certain words we carve up reality in a particular way.

These nuances in conceptualization and vocabulary cannot of course be separated from sometimes deep differences on other levels. They affect the way in which research and educational systems are organized in the countries concerned. They bear on the kind of relationships between researchers and society. they are revealed in the very unequal economic power of the media. We find them expressed in the original traditions and materials of literature, cinema, or the performing arts. The policies for culture in different countries embody them, as do the relationships between citizens and the political establishment, etc.

The analysis by Jan Nolin, for example, of the way in which the demand for democracy has shaped a role for scientific information, well reflects the 'participative' concept of democracy current in Sweden, and more broadly in Scandinavian countries. The articles by James Bradburne on the one hand and by Yakov Rabkin and Elena Mirskaya on the other illuminate how the situations in the countries of Eastern Europe and the republics which have emerged after the collapse of the Soviet Union are connected to the specificities of the recent histories of these countries, and of the changes which they are undergoing today.

By superimposing these differences we are better able to understand the contrasting nature of the situation in Europe in areas such as scientific information and communication. The picture painted of the press by Pierre Fayard is particularly striking in this regard. We could extend it in several directions. We could show how the professionalism of British television, in conjunction with the important resources at its disposal, has led to the production of films which are appreciated throughout Europe (the French and German approaches are less successful in other countries). We could also develop the final part of the article by Marcel van den Broecke and show the uniqueness of the initiatives taken in the Netherlands to grant citizens immediate access to information ('science by phone', 'science shops', etc.). And we could draw attention to real differences behind apparent identities.

Britain and France, for example, are noteworthy for the richness of their production of original books, translations of which are the major works on the market in other countries. Yet while most of these are written in the case of Britain by professional science writers, in France they are written by the researchers themselves. Formally similar initiatives can, from another point of view, retain a strong national imprint. This is notably the case for the different 'science weeks', each of which has its own personality: the Italian Settimana della Cultura Scientifica is an open-door event held throughout the country, while the Science en Fête is carried out in a spirit very similar to that of the other fêtes organized in other areas of France. And like the annual congress of the AAAS, the British Association's Science Festival is an occasion for researchers to report original results.

All of this of course does not mean that European countries are not confronted with identical problems, and share several concerns. All are seeking the best way to improve scientific education with a view to ensuring the production of future generations of researchers and engineers. To generate a genuine democratic debate on scientific and technological choice they are all seeking to raise the level of knowledge and understanding of science by the public.[1] It is interesting to see on this point the way in which the remarks by John Durant or Jan Nolin converge with those of John Krige in his analysis of how CERN is trying to justify socially its existence and its activities. Far from any idealized image, it is science as it really is, including its institutional and social aspects, which have to be understood – the scientific system as it really functions, with its enormous efficiency but also with its weaknesses and its limits.

As anxious as they might be to conserve their identity, European citizens also recognise that they are united by what Edgar Morin has rightly called 'a collective destiny'. They know and they feel that multiple historical ties exist between their countries–ties which give a meaning to an expression like 'European culture' (in the second sense of the word).

The multiplication of collaborative efforts

For several years in the context of a general increase in transnational cooperation in Europe, people and institutions active in the fields of information, culture and scientific education have drawn closer to each other. Some collaborative projects have taken place on the European scale, or have been proposed. Some have been mentioned in this volume: the ECSITE initiative for science centres by Melanie Quin, for example. Several colloquia have also been organized, bringing together professionals from different European countries – scientific journalists, managers of museums, people concerned with public relations in research organizations, etc. People have got to know each other better, new linkages have been established and networks have been gradually put in place. The Commission of the European Communities was associated with several of these initiatives. The promotion of information and of scientific culture in Europe appears indeed naturally as one of the tasks to be undertaken by the Community along with its more specific research programmes.

With the launch, at the initiative of Commissioner Antonio Ruberti, of the 'European Week for Scientific Culture' actions which until now have been somewhat restricted have undergone an order-of-magnitude change and moved to a rather broader and more significant plane. Operations of this kind carried out at the national level have shown clearly the impact of the simultaneous organization of events concerned with communication, and held in different places. In a similar spirit the objective of European Week for Scientific Culture is twofold: to alert above all political decision-makers and European citizens to the important issues at stake in scientific culture and education in Europe; and, simultaneously, to make the public aware of science in its European dimension. This has two aspects. Firstly, there is European scientific cooperation as it existed in the past, and as it is carried out today by CERN for nuclear physics, EMBO for molecular biology, ESO for astronomy and the European Community. Secondly, there is science as it is practised, perceived, and displayed in other European countries – other people, other traditions, other views of science and its relationships with society.

During the European Week for Scientific Culture some twenty events will be organized–events which are European in their theme ('European Science', 'Science in Europe'), in the public at which they are aimed (necessary in several countries), and in their modes of execution (with transnational collaborations). A key feature of the initiative is the wish to bring together professionals of information and scientific culture with those in educational milieux. In most European countries these two worlds tend actually to ignore one another. It is not even unusual to hear specialists of scientific culture presenting their work as intended to compensate for the deficiencies of the educational system. In fact as many have remarked, including John Ziman in his contribution to this volume, there can be no real understanding of science without a sound educational base. To improve this, those responsible for educational systems could benefit from a better understanding of one another and from a closer contact with professionals dealing with scientific information.

Scientific cultures in Europe

In everything that has been said up until this point, it has been scientific culture in the first sense of the term which has been under discussion. How are we to pose the problem of the relationships between science and culture in the two other usages of the word? The place and the weight of science in European culture in the most objective sense of the term certainly depends on factors other than policies and initiatives of a cultural and educational type. To a large extent they depend on the importance given to research by governments and by industry, and on the degree to which these take seriously the statement that our societies – to use Tullio Regge's expression – are 'condemned to do research'. In fact, as Antonio Ruberti has stressed, Europe cannot hope to be a producer of new knowledges unless the countries of which it is composed coordinate, in a manner far more determined and profound than they have done up to now, their policies and their investments in scientific matters.

One can also measure the importance of science in the culture of a country or of a civilization by the degree to which the growth of scientific knowledge, far from being carried out blindly, is an object for reflection. Here too it seems necessary to study the issues at the European level. The aim of the conference 'Pensées et pratique de la science en Europe' – 'Theory and Practice of Science in Europe' – organized by Dominque Lecourt, for example, is both to confirm the existence of a genuinely European conception of science and to look at the conditions under which a study of these questions could be organized on our continent.

Along similar lines, and with the impetus provided by Antonio Ruberti, the Commission hopes that by launching a series of specific studies, conferences and publications, it can stimulate reflection and debate on science in Europe. Here as elsewhere the existence on our continent of a large variety of ways and of traditions of thought undoubtedly constitutes an advantage.

With this remark we touch on questions associated with the anthropological meaning of the word culture. In this third sense it is obviously convenient to speak not of 'a' but of 'the' European cultures. European cultural diversity shows notably in the way in which research is organized and science taught in different countries, and is expressed in the way in which science is viewed and considered. Cultural differences emerge too in specific modes of innovation, in particular schemes of production and diffusion technologies, and so on.

It is not by chance that the Community, in parallel with increasingly well-supported steps to promote education and scientific culture, is increasingly led to take initiatives intended to reflect more deeply on science and on technology, and to study their cultural components. The deep meaning of the Treaty for European Union (the Maastricht Treaty) is to give the Community a character which is broader than that of a mere economic union. To a lesser degree of course than in foreign policy and security affairs, the new Treaty, while explicitly placing research policy at the service of all other policies, extends the mandate of the Community into the fields of education and of culture. At the intersection of these three domains there is thus a vision of science which is richer and larger and which the Community is being encouraged to promote.

During the debates on the Maastricht Treaty it emerged quite clearly that there was no question of building a united Europe at the expense of the specificity of the patterns of life, the behaviour, and the views of the world which characterize the countries of which it is composed. This then, in truth, is the real issue at stake in the construction of Europe: far from destroying them, how are we to develop and take advantage of the differences between regions and cultures which constitute the uniqueness and the richness of our continent?

From this point of view the actions which will be undertaken by the Community concerning questions of science, culture and education are an important test case. They will allow us to measure the degree to which Europe, by means of cooperative actions, can integrate and overcome the tension between identities and differences which have shaped its history. By presenting to us a number of different aspects of the relationships between science and culture in the various senses of that word, this volume constitutes in any case an excellent illustration of the way in which Europeans can, by improving their knowledge and understanding of one another, mutually enrich each other with their differences.

Reference

1   The danger, as Jean-Marc Lévy-Leblond has stressed elsewhere, is to arrive at a situation in which one makes the possession of such knowledge a necessary condition for any right to criticise scientific and technological matters. After all, as he says, 'we don't demand of citizens that they have a knowledge of constitutional theory before we allow them to vote, and the juries of the Cour d'Assises are not required to produce a certificate showing their aptitude in criminal law before they are consulted.' See Lévy-Leblond, J.-M., 1992, En méconnaissance de cause. Qui a peur de la Démocratie? *Le genre humain*, November.

Places

Science in post-Soviet Russian culture

Yakov M. Rabkin and Elena Z. Mirskaya

Soviet science in Russia and beyond

The Soviet scientific enterprise was the largest in the world. Now that it has collapsed, it remains to be seen whether the scientific community of the former USSR can preserve the cultural and cognitive links fostered during the history of the Soviet Union, and whether the culture of science is resilient enough to transcend the new political borders. Most national sciences in the Soviet Union were deliberately established by Moscow as a token of the republics' national prestige. Both science and the semblance of science were consciously promoted by the Communist partocracy, in Moscow and in the periphery.

Most peripheral sciences in the Soviet republics had an important ornamental, propagandistic function reflecting their political origins. National academies of science were organized on the basis of field offices of the Soviet Academy of Sciences, most of which had been upgraded to fully fledged status by the end of World War II.[1] Even though technically autonomous, the academies were heavily dependent on Moscow, and had few horizontal links with sciences in the neighbouring national republics. Significantly, Russia was the only republic not to 'deserve' its own academy. Since Moscow was clearly the recognized centre, Russia's science needed no internal symbol of power and prestige. It was only the decomposition of the Union that led to the establishment of Russia's academy which, quite deservedly, promptly declared itself the successor to the Soviet Academy. Thus there are at least two sets of issues in post-Soviet science. One set has to do with the various national sciences which may have to maintain links with science in Russia, to restructure its links with world science which used to be mediated by Moscow, and to establish new links with scientific institutions in its geocultural habitat (Turkey, Iran, India for the southern republics, and Central Europe for Ukraine and Belarus). Science in the centre faces another set of issues. With a few exceptions, the secession of national sciences is likely to have negligible effect on the conduct of research in Russia's institutes and laboratories. What mainly concerns Russia's basic science is the precipitous decline of its status as a key priority for the State. This has immediate repercussions in terms of research funding, recruitment of new scientists, and – even though political restrictions are no longer in place – its links with world science. All this impacts on the unique culture of science which developed in the course of nearly three centuries of the history of Western-style science in Russia.

We might recall the way Leo Tolstoy began his *Anna Karenina*: 'All happy families look alike. But every unhappy family is unhappy in its own way'. The situation of science in every republic of the former Union must be treated in its own right. This article will focus on the fate of Russia's science as a distinctive culture and on the prospects of its integration into world science.

Cultures of Soviet science

Several cultures of science developed in the Soviet Union, ranging between two extremes. At one extreme there was a strong motivation to do research for both intrinsic and extrinsic reasons: to make a mark on world science and to contribute to the welfare of the country. At the other extreme was a culture based on an interest in science as a source of prestige and financial benefits, but not in research activity per se. Research was routinely carried out by subordinates and then attributed to administrators of research institutes and other members of the ruling Communist class, the *nomenklatura*, who were often entirely unconnected with science. A scientific degree offered both symbolic and material capital as well as a relatively secure place of refuge for members of the *nomenklatura* wishing to hedge their political bets.

Soviet science was hierarchically structured and exhibited little incompatibility with the prevailing totalitarian system. Scientific workers situated at different levels of the pyramid led different lives and had different goals.[2] The top of the hierarchy, essentially the scientific echelon of the *nomenklatura*, behaved like the rest of the Soviet ruling class. The domination of the governing elite led to a significant polarization of cultures within Soviet science. It has even been observed that 'scientific researchers became marginalized not only from society but also from their own social institution – science'.[3] To a large extent, this was due to the decisive influence of the State on the social relations of science.

The State literally imported Western science from Europe in the early eighteenth century and has remained its exclusive patron ever since. Beyond strategic considerations, the cultivation of science made a persuasive claim on behalf of Russia that she belonged to 'the civilized Europe'. However, the 'low culture' of much of Russia's population remained little affected by these occasional bouts of modernization, which invariably befell Russia from the top down, i.e. from its political and intellectual elites operating under government control.

Science, an activity historically linked to Western liberal values (see, for example, the charter of the Royal Society of London), suffered a peculiar fate in Russia's collective psyche. Attitudes to science often served as a reliable litmus test to tell a 'slavophile' from a 'Westernizer'.[4] Long before science-based ethics was used to justify massacres in the twentieth century, the slavophiles were attacking science as foreign and therefore spiritually evil for Russia. The Westernizers saw in science a sign of advanced, moral, European civilization. For them, science, just like Russian literature, has been elevated to be more than simply science: a religion – a source of moral values and societal responsibility. But neither attitude paid attention to the actual roles of science in Russian society.

The communists proclaimed science to be the foundation of a new social order and attempted to elevate it to the role of a moral substitute for organized religion. Even though in the mind of the masses, to whom the communists paid so much lip service, scientists were largely associated with an imposed alien order which gave scientists privileges and resources well above their perceived usefulness for society, the State remained a potent protector of science, both materially and ideologically. This intimate – and, in the case of physics and mathematics, unflinching – support of science by the State enabled scientists to assume a degree of corporate autonomy unparalleled in other Soviet professions.

At the same time the State never trusted the scientists. A certain anti-intellectual and anti-scientist feeling was maintained in the masses by means of the party-controlled press which would regularly print reports of misdeeds by individual scientists, alongside laudatory prose in praise of Soviet science.

While the people may mistrust the scientists, the scientists feel quite differently about the people. Russia's scientists are a part of an intelligentsia. They share its specific culture, including a traditionally high degree of responsibility for the welfare of 'the people'. Soviet propaganda cultivated this sense of responsibility, and made scientists feel guilty for the relatively good life provided for them by the State. This guilt easily explains the devotion many scientists feel to the people, and is part of a culture which puts emphasis on selfless devotion to both science and society. The life of Andrei Sakharov, admittedly an exceptional personality who epitomizes one extreme in the range of scientific cultures outlined above, succinctly illustrates this combination.

Science, because of its elevated status in Soviet ideology, also offered an acceptably neutral vehicle for civic discourse. Political issues had to be clad in scientific garb in order to make them appear objective and incontestable. Today there is no longer any need to disguise political questions in order to debate them as issues of science, and the new refreshingly free political process has made scientific credentials largely irrelevant to decision making.

Science has now been left to its own devices. Political elites have turned away from science and look instead to popular support as a source of political legitimacy. The State no longer needs science to cultivate illusions of a better future. Post-Soviet elites view science as largely useless: privatization is now used to maintain a belief in a better future. Incomplete assimilation of science by Russian society becomes a crucial liability when it goes through another bout of enforced modernization. Anti-Western and, consequently, anti-science feelings become sharpened and adversely affect the image of science in today's Russia. The cultures of science, and the self-images of scientists, undergo a similarly rapid change.

Science, scientists and change

The role of intellectuals – in Russian *uchenye*, literally 'learned', stands for both scientists and scholars – in the on-going changes in Russia may appear primordial. Scientists as an occupational group not only embraced *perestroika* more enthusiastically than others, but some of them had played a conspicuous role in promoting this change. Many well-known scientists who had long been exposed to Western, mainly American, science transmitted its influence initially into the culture of Soviet science, and later into the realms of political and economic culture.

The Communist Party used to associate projects of economic reforms with the names of prominent *uchenye*. The partocrats, with the notable exception of Gorbachev, preferred to remain nameless in the process of reforms which the Central Committee initiated. For example, three decades ago it was not Kosygin, the Prime Minister, whose name was associated with attempted but ultimately ineffective industrial reforms, but Liberman, a *uchenyi* and – and this is often synonymous in Russian popular consciousness – a Jew. The government of Yeltsin has more *uchenye* among the ministers than any previous government of the country. Increasingly *uchenye* are exposed as a convenient scapegoat for Russia's problems, be they economic, ethnic or environmental. Yet the public largely attributes the country's difficulties to politicians rather than to scientists.

The Chernobyl nuclear catastrophe contributed heavily to the decline of the prestige of Soviet science. *Glasnost*, while hailed by most scientists, also became a potent instrument for dethroning science. The shrinking of the Aral Sea, and hundreds of other ecological disasters, came to be regularly reported in the local and national press, often by scientists themselves.[5] With a growing frequency, science was portrayed as a menacing monster, an utterly novel image for Soviet citizens to absorb.

At the same time, *perestroika* opened the gates to a variety of alternatives to the monopoly of science as the source of truth. Daily horoscopes, advertisements of healers, interviews with witches, all of it previously unknown in the Soviet media, erupted into the post-Soviet public arena. While a coherent anti-science discourse is yet to emerge from these disparate phenomena, several scientists have reacted with anger and fear. They see in the emergence of these phenomena a serious danger to science.[6] In fact, it may be a more serious danger to scientism as a system of beliefs than to science as a research activity. Thus, in a conversation with one of the authors, a prominent cleric of the Russian Orthodox Church said that he saw an important role for science in the new society once it had given up the arrogance of omniscience and claims to the role of moral guide. 'If science can accept as its basis the primacy of spiritual and ethical values, and will be able to combine the material and the spiritual, it will have a good prospect for the future.'

The current crisis in Russia's science has no parallels in world history. While the October revolution of 1917 initially threatened the continuity of science, several factors helped science survive and prosper under the communists. Firstly scientism, as a constituent element of the new ideology, ensured an ideological commitment of the communist elites to science. Secondly, the reinforcement of a strong centralized state obsessed with military concerns led to an impressive expansion of the actual research system under the Soviet regime. Public perceptions of science now ebb and flow in many countries. In industrial democracies, the consensus about the basic need for science survives swings in public opinion. In developing countries, where science remains an alien activity for the masses, it is known to be more vulnerable to changes in political orientation.

A crisis of science and, more generally, of rationality also occurred in post-Versailles Germany. It has been shown that it ended in an explosion of new scientific ideas, offering a different, less deterministic world view.[7] It remains to be seen if the current crisis of post-Soviet science may affect scientific thinking in a comparable manner.

Today's crisis undermines the cultural foundations of science. After decades of pro-science propaganda science is suddenly no longer viewed as a useful societal activity. Its excessively generous development under the Soviets, against the background of relative estrangement of science from Russian national traditions, generates cultural resentment among people across the political spectrum. Popular support for science is limited at best. Post-Soviet science has lost most of its extrinsic value for military and prestige purposes, and has yet to acquire an intrinsic value comparable to that of the performing or fine arts. Addiction to exclusive state support harmed Soviet science in different ways. It became inept in mobilizing public support, while the public came to resent the formerly privileged position of science in Soviet society.

According to a recent survey on public attitudes to science conducted by the sociologist V. A. Yadov, only 8 per cent of the respondents believed that 'science did more harm than good', while 52 per cent saw in science more good than harm. In ascending order, the perceived harm done by different sciences appears as follows: medical sciences, biology, physics, humanities, social sciences and, finally, chemistry – the most maligned of the sciences. Most respondents expect science to yield immediately useful results, while only 14 per cent support basic research. Not suprisingly, the least educated strata exhibit less enthusiastic attitudes to science.

The decline in the social status of science leads to a constant outflow of personnel. The outflow consists mainly of ambitious mobile men, who are less than totally committed to science as the principal goal in their lives. While this outflow, along with other critical developments in the post-Soviet science, may actually strengthen the classical values of science among the remaining few, culturally aberrant phenomena have begun to pollute post-Soviet science. Science has come to be involved in criminal activities. Narcotics and illegal blood banks and organ transplants are some of the options openly discussed in the science news press.[8] This marks a drastic departure from the old cultures of science, a departure well beyond the range of scientific cultures outlined above.

For those who stay in science, a novel phenomenon is a growing involvement in science management. Scientists have become interested in the actual running of science as a social activity. Russia's scientists are gradually emancipating themselves from the exclusive domination by the State and are getting organized into a self-styled scientific community. New periodicals, both printed (*Poisk* and *Nauka i biznes,* formerly *Radikal*) and electronic (*Kurrier RAN*), have sprung up to inform scientists about the politics of science, and to provide a forum for debating science-related issues. They form an essential element of the new more participatory culture of Russia's science. This outburst of free communication among scientists takes place against the background of a numerical decline of popular science publications which have yet to find private alternatives to the generous financing they used to enjoy under the communists.

Currently, Russia's science integrates distinctive elements of the Anglo-Saxon scientific culture such as the emphasis on horizontal organization of scientists as a community. Perhaps as a reaction against the old hierarchical structure of science, the Russian scientific community today seems to have a more developed system of internal communication than, for example, France's scientific community which lacks a functional equivalent of London's *New Scientist* or Moscow's *Kurrier RAN*.

Russia's scientists are willing to experiment with new, more flexible forms of organization of science which, once again, is characteristic of science in Britain and the United States. It is these countries, rather than France or Germany, which have become the usual source of inspiration for innovations in the ethos and organization of post-Soviet science. Science fits the general post-Soviet trend of admiration for things American which, in this case, is further strengthened by the role of English as the dominant lingua franca of science.

Science in major continental countries offers more centralized patterns of organization. These patterns are historically more compatible with the tradition of Russian and Soviet science. This compatibility leads European efforts to intensify contacts with Russia's science towards the old centralized structures and the familiar *apparatchiks* of Soviet science. For example, it is quite natural for France's CNRS, whose very formation was apparently inspired by Soviet experience, to deal with central bodies such as the Academy of Sciences or the Ministry of Science in Moscow, rather than with new independent structures whose identity, reliability and exact configuration in the former USSR need to be ascertained in every single case. Cultural affinity and institutional inertia are likely to lead much of Europe's assistance to inhibit rather than promote change in post-Soviet science.

The culture of post-Soviet science is in flux and may be more sensitive to foreign influence than other varieties of culture. In the absence of a tangible Japanese presence in Russian science, and with the cultural limitations so far experienced by Europe's scientific assistance, it is the American model that is likely to leave the most significant mark on the evolving culture of science in the former Soviet Union. Post-Soviet science has opened up to the world, and its future largely depends on the nature and degree of the contacts it develops in the West.

Acknowledgement

We are grateful for advice on an earlier version of this article from Spas Spasov and Adel Ziadat.

References

1   Rabkin, Y.M., 1992, Academies: Soviet Union. *Encyclopedia of Higher Education* (New York: Pergamon Press), pp.1049–1055.

2   Yaroshevsky, M.G., 1991, Stalinizm i sud'by sovietskoi nauki. *Repressirovannaya nauka* (Leningrad: Nauka), pp.9–32.

3   Mirskaya, E.Z., 1992, The problem of justice in Soviet science. Paper for the workshop 'Science and Technology with a Human Face', Moscow, September 1992, p.13.

4   Vucinich, A., 1970, *Science in Russian Culture* (Stanford, CA: Stanford University Press), Vol. 2.

5   Yanshin, A.L., and Melua, A.I., *1991, Uroki ekologicheskikh proschetov* (Moscow: Mysl).

6   A good example of this reaction was offered by Sergei Kapitza, a well-known physicist and popularizer of science, who vehemently denounced these novel alternatives to science in the course of a seminar on anti-science movements in the USA and the USSR held at the MIT in May 1991.

7   Forman, P., 1971, Weimar culture, causality, and the quantum theory, 1918–1927: adaptation by German physicists and mathematicians to a hostile intellectual environment. *Historical Studies in the Physical Sciences*, **3**, 1–115.

8   Editorial, 1992, *Radikal*, **18**(75), 10.

Authors

Yakov M. Rabkin is Professor of the History of Science at the University of Montreal, CP 6128, Succ. A, Montréal, H3C 3J7, Québec, Canada. He has devoted a book and several articles to cultural and political aspects of Soviet and post-Soviet science. Elena Z. Mirskaya heads the Department of Sociology of Science at the Institute for the History of Science and Technology, 103012 Moscow, Staropanskii per. 1/5, Russia. She has published and edited several monographs and articles on issues of science in society.

Scientific culture to the letter

Bernard Maitte and Jean-Marc Lévy-Leblond

On 30 February 1993 a Parisian weekly published an extensive survey on the development of scientific and technological culture in France. It subsequently received an immense amount of correspondence from its readers, which to date has remained unpublished. We are pleased to have this opportunity to present a few extracts which illustrate pretty well the current situation in France.

From Mme A, lecturer in French literature

So, we now have a 'culture of science', do we? Well, why not, since nowadays everything is scientific (opinion polls, beauty creams and theology) and cultural (rock music, cookery and advertising). Yet I would rather have thought that the strength of science lay in finding an answer to the question 'how?', while the role of culture was to ask the question 'why?' Would it not be better to let science work at resolving its problems, and culture to try to make sense of other issues? What can be gained from merging the two? Not long ago I attended an exhibition called 'The Dance of Matter', which was designed to initiate the layperson into particle physics and basic interactions by presenting display panels on which semi-popularized texts with literary pretensions were accompanied by reproductions of modern paintings. I didn't understand quantum theory any the better, and was saddened to see works of art reduced to a purely illustrative role. How can the singularity and the fortuitous nature of culture, with its unique products, live in harmony with the normality and necessity of science and its collective work? This is not to belittle science – Victor Hugo's 'splendid servant' – but on the contrary to recognize its specific character in the ensemble of the arts and professions, rather than to treat it as a cultural genre (or to think of it as the nth art, after the eighth or ninth etc., where n equals infinity). And it does not help culture, which is already threatened by technology and commercialization, to force it to integrate itself into a dimension of human activity that is so alien to it. It is precisely this singularity which must be preserved. If, as some scientists claim with pretentious naivety and some artists accept with obsequious cunning, science and art 'are convergent', 'share the same goal', 'express the same creative urge' and other pathetic sophisms, they will both lose that which makes them great. The fact remains that difference does not necessarily lead to indifference. It is precisely because science and culture do not follow parallel paths that they are able to meet! Such 'brief encounters', fleeting and unexpected though they may be, are certainly more fruitful than any organized institutional alliance, perhaps precisely because of the feeling of incompleteness that they leave behind them. In any case, you only have to be able to read, listen or look to realize that culture has not waited for an invitation to concern itself with science, as with all other branches of human activity. For example, great works of theatre like Bertolt Brecht's *Life of Galileo* (and its unforgettable recent production by Antoine Vitez), works of literature like Umberto Eco's *The Name of the Rose*, the sculptures of Piotr Kowalski, Xenakis' musical scores, and Alain Resnais' cinematic masterpiece Providence – all these tell us more about science, its struggles, its promises and its threats than any popularized scientific exhibition.

From M. B, director of a Centre for Scientific and Technological Culture

Thank you for your article on ten years of scientific and technological culture in France. Since you only make allusions – albeit many and favourable allusions – to the activities of a regional Centre for Scientific and Technological Culture, might I be permitted to explain what these actually are? Our goal is to throw open to discussion those developments in society in which science and technology are today playing an increasingly important role, and to be a meeting place where science, technology and the arts can confront one another.

We put together travelling exhibitions, light-weight and easy-to-use 'exploration cases' for schools, clubs, etc. By doing this we seek to awaken the pleasure of discovery and understanding, to rediscover in science an attitude of questioning and wonder, and to develop the motivation that precedes and stimulates every step towards knowledge. We develop activities that are aimed at young people: as a result of our efforts, clubs are being set up all over the region which enable their members to carry out experiments, to make science themselves, to discover that science and technology proceed by trial and error. We arrange 'research passports' that enable young people and their teachers to discover science as it is happening and can be seen in the laboratories and companies in our region.

Those in positions of authority in education and culture come to us to discover how to organize projects in their own towns and districts. We mediate between project leaders, representatives from industry and researchers. Our training programmes introduce teachers and organizers to new methods, and show them how to hold debates and discussions on current problems relating to science. By means of 'news rooms' installed in the towns of our region, which are updated every three months, we treat in depth, but in an attractive manner, the social implications of scientific development. We also put together documents on these subjects for schools, industry and the media. Finally, our centre for multimedia documentation publishes guides, directories and catalogues. Together with the region's other centres of documentation – namely the libraries – it supplies books for young people, fosters our productions by putting together documentary dossiers, and enhances the exhibitions and debates that we organize by providing reading corners and videos.

We do not conduct our activities in isolation: about ten centres exist in France, either already operational or in preparation. Their intention is to make clear the full depth of the sciences, to place the sciences in the context of their history and the development of our societies, to put them in perspective, to write them into the changes through which we are living, and to enable our citizens to understand and master current developments.

From M. C, shop steward

What can the idea of scientific and technological culture mean for a population that is primarily preoccupied with the economic decline of its region? Its economy rested on three pillars that appeared solid during the 1970s: the textile industry, the iron and steel industry, and coal mining. All three have collapsed. Following their disappearance, a project was developed to create places by which to remember them – known as 'ecomuseums' – to preserve the traces of a defunct economic activity and technological culture. Such initiatives could not redevelop industrial sites, but have they at least been able to preserve the industrial skills that could and should be transferred to the new technologies? For this to have been done, it would have been necessary to link this policy with measures for preventing the anticipated forced retirement of workers with years of experience behind them. This was not done. Cultural policies are still kept separate from economic policies. Those in authority believed that the new technologies would provide jobs. They engaged in a policy of inciting consumer demand and then tried to pre-empt it by launching educational initiatives, the most famous of which was 'Computers For All'. These initiatives faltered at the twin stumbling blocks of insufficient initial training of people in the workplace and inadequate schooling of the young, not to mention the prevailing economic constraints. Don't you think that we should be wary of illusions? Museums of technology were founded in the nineteenth century during the World Exposition movement. Museums of natural science were founded by learned societies. The Palais de la Découverte (Palace of Discovery) was built by the Popular Front. All fell into disuse once the necessities that led to their creation became less of an issue. Won't the Centres for Scientific and Technological Culture suffer the same fate? They were born out of a generous idea, but won't it take too much time for this idea to be realized, the time that it takes for all sectors of society to be affected by their actions? And will the authorities agree to underwrite this policy in the long term?

From Mlle. D, artist

As an art school teacher and video maker, I read your investigation into scientific and technological culture with great interest, an interest which I share with many other artists. Some seek in the new technologies (information technology, holography, video, memoryform materials, etc.) novel means of expression. Others are interested in the conceptual upheavals which quantum physics and molecular biology bring to our understanding of the world, and try to reflect them. Beyond certain inevitable effects of fashion (such as the vogue for fractal images and their technokitsch aesthetic) there is, in my opinion, a real desire for an opening up aₙd exchange of ideas between science and culture. Eloquent examples of this include the fantastic exhibition staged by Louis Bec in Avignon in 1984 entitled 'Le Vivant et l'Artificiel' ('The Living and the Artificial'); 'Electra', an exhibition of contemporary art in Paris organized by Jean-Christophe Bailly during the same period, and, a little later, 'Les Immatériaux' ('The immaterials') at Beaubourg. During the 'Arborigènes', a collaboration between the sculptor Ernest Pignon-Ernest and biologist Claude Gudin, strange plastic sculptures dressed in seaweed invaded many parks and gardens. Centres for Scientific and Technological Culture open their doors to contemporary artists. For example, the centre at Thionville exhibited Denis Pondruel's impressive installations. The journal *Alliage* regularly features painters, sculptors and photographers in its pages. The Institute of Astrophysics in Paris was very successful in organizing regular meetings between invited researchers and artists – painters, comedians and writers. Conversely, it is now not unusual for our colleges of art or other cultural institutions to invite physicists, mathematicians and biologists to come and present and discuss their work. These meetings are often frustrating – the scientific ignorance of the artists is matched only by the scientists' lack of artistic culture (with only one difference: the former recognize their limitations more readily than the latter) – but they are always valuable experiences, even though there is no expectation of a miraculous convergence, and one is always wary of any illusory syncretism.

It is a matter of some regret that the Ministry of Culture has distanced itself somewhat from this movement. After declaring its interest at the beginning of the 1980s and taken various initiatives (for example, the creation of a National Council for Scientific Culture in 1985, the increased interest taken by the National Centre of Literature in 'scientific literature', etc.) this ministry, while it undertook bold actions in so many new sectors, has been unable to demonstrate the same originality here. Is it that science is more intimidating than rock music or fashion? One could have hoped for a more inspiring or even spectacular policy: for example, state commissions or competitions. There would be no shortage of ideas: a 'Monument to the Unknown Animal' (guinea pig? fruit fly? bacterium?) to be erected in the gardens of the Pasteur Institute; an oratorio on the construction of the atomic bomb (the Manhattan Project) – a dramatic subject if ever there was one; a musical comedy on the life of Einstein (to be called 'Einstein on the Roof' perhaps?); a huge Foucault's Pendulum to be erected outside the new museum of arts and crafts; a 'son et lumière' spectacle at the Nançay radio telescope (in the style of the inauguration of the Olympic games at Albertville?). But when all is said and done, why shouldn't patronage take over? Large technological and scientific companies, both public and private, finance festivals of classical music and theatre seasons. So why can't they also help to foster cultural activities that concern them directly?

From Mme E, school teacher

You devoted a considerable part of your last issue to the activities of the Regional Centres for Scientific and Technological Culture. I would like to describe an initiative that I was able to carry out with the help of one of these centres, which made available to us one of its 'exploration cases' dealing with the concept of symmetry. A cube enclosing its secrets intrigued me from the start. During the training I was given I saw it open out into mirrors, games, anamorphoses, rugs and cards. Here, marvellous fragile membranes of soap solution stretched between the edges of a solid. There, palindromes were formed. Music by Bach imposed upon us its rigour, while the tryptich of the Mystical Lamb was reconstructed. Quartz and calcite demonstrated the variety of their forms, while a model helped to explain them. Architecture, sculpture, history of science and epistemology all combined to provide a feast for my students, who were enthusiastic about making discoveries, eager to have hands-on experience, delighted at seeing the arts, literature, music, physics and mathematics in juxtaposition, and were the richer for seeing relationships that enhanced their understanding being woven between disciplines that are too frequently taught in isolation. The adventure continued throughout the year with visits, surveys, discoveries, debates and initiatives carried out by several teachers in conjunction with the Centre for Scientific Culture. The result was conclusive. The school benefitted greatly from the experience. What is more, I must tell you about a spectacular conversion. In our school the relations between the science teachers and the documentation centre were cordial, but did not exist at all on the professional level. I would say that the responsibility for this lay on both sides. I showed the 'exploration case' to the school's information resources manager, who was knocked out by it. A little later he told me that he was attending a training course on scientific books organized by the Centre for Scientific Culture. He returned from this course cured of his complexes, and has been able to rethink his criteria for analysing books, which he previously could not apply to scientific books because he had considered himself unqualified in the field. Since then, there have been fruitful exchanges between the Centre and the school. We have staged workshops at which we have analysed scientific documents. These often finish with the presentation of a number of works of fiction that initiate lively debates on the relationship between science and society.

One last point: the French teachers are also being asked by their students who have been attending the analysis workshops to continue the discussions in class.

From Mme F, librarian

The enthusiasm for new forms of making science 'cultural' must not lead us to forget that which remains, in my opinion at least, at the heart of all cultures. Centres for Scientific Culture, museums and television programmes are all very well, but books are the most important of all! We are not about to leave the Gutenberg galaxy. Books are the best instrument of culture and will remain so, because they allow a personal, controlled and open approach. They alone allow one the time to reflect and assimilate, to return to a previous page (or skip on ahead), and prompt one to digress and consult other books. Therefore your assessment of the development of scientific culture in France is very incomplete, because it fails to stress the remarkable expansion of scientific literature aimed at the general public that has taken place over the last twenty years. Our country has a long and rich tradition of popularizing science, notably in the eighteenth century by Abbé Pluche and Fontenelle, and by Flammarion, Figuier, etc. in the nineteenth century. However, this tradition gradually fell into decline, so that by the 1950s and 60s it had lost contact with active scientific research, which was now highly specialized and had little concern for its public image. It took the ideological shake-ups of 1968 to give rise to a dynamic editorial sector based on a more critical and open notion of science. Over the last 20 years many large publishing houses led by le Seuil, and followed by Fayard, Odile Jacob and others, have produced new and original collections. The traditional concept of 'popularization' is frequently superseded by a less paternalistic approach, attaching as much importance to the reader's questions as to the author's answers. Science was challenged – by philosophy, history, economics and politics. The importance and originality of this editorial policy in France are confirmed by the response it has received abroad, as testified by the large number of published translations. The public authorities have played their part in these developments, for example by creating a committee for 'scientific literature' at the National Centre for Literature (under the auspices of the Ministry of Culture), which provides funding for the publishing and translation of such works. Many book fairs now invite scientific authors to display their books alongside writers of fiction – and this even happens on television programmes dealing with literature. So, long live scientific literature! For as Galileo himself wrote, 'Is not Nature a great book eternally open to our gaze'?

PS. May I add to my eulogy on the book as a tool of scientific culture, a eulogy to the radio? In France we can benefit from France-Culture – an unparalleled source of thought-provoking and stimulating information covering all aspects of culture including science, whether in the form of in-depth or magazine programmes, dealing both with science's ancient history and its recent discoveries. In what other country can one listen at ten o'clock in the morning to a programme on the birth of mathematics in physics (from Newton to Varignon); at lunchtime to France-Culture's 'Panorama', featuring a no-holds barred debate on the recent film about the work of Stephen Hawking (*A Brief History of Time*); in the afternoon to an update on the most recent theories concerning comets, and in the evening to a famous biologist talking about his love of music. I can vouch for the fact that these programmes are by no means reserved for the intellectual élite: many young people (school children and students) or not so young people (cultural representatives on company committees, senior technicians, teachers and the retired) come to our libraries looking for books that give more information on the subjects covered by these programmes. Let us not forget the more unobtrusive, but nonetheless real, presence of science on general public channels such as France-Inter and France-Info. If television serves science badly, isn't this because, when it comes down to it, we still do not know how to make real television? All the more reason for not underestimating or devaluing those cultural tools which we do know how to use: the book and the radio.

From M. G, emeritus university professor

I feel I must give vent to the indignation that I felt on reading your last issue. Anyone who has any real scientific knowledge cannot fail to be outraged by the pathetic entertainments put on for public consumption by establishments that are supposed to represent science. I went to la Villette: the Géode there is the ideal device for anyone who wants to take a white-knuckle ride in an armchair – we're in an aeroplane that is taking a nose dive... we're going to crash... everyone in the room ducks! The horizon pulls away and the commentary tells us that 'Galileo declared the world to be round'. We go the planetarium, where we see the mast of Christopher Columbus' Caravelle pitching and tossing under the stars, and emerge drunk on images and sounds.

This tyranny must end! Intellectual dishonesty supports all other forms of dishonesty, and these prestigious creations require large amounts of money forcibly taken from meagre research funding that is already overstretched. Our laboratories need to survive – in peace and quiet, without opening their doors to all and sundry. They should not be deprived of the ability to recruit staff for two years so that la Villette can be opened to the public. Your 'scientific culture' is nothing but fashionable popularization, and bad popularization at that. It takes continuous renewed effort to rise to the level of science: science is a merit which can be achieved and which leads to the truth, reason, and the sharing of the results of research. Only this latter leads to progress – progress which benefits the whole of humanity – and allows us to reject the irrational currents of thought which throughout history have hampered man's development and which today are making a dangerous reappearance. Let us stop harping on about ethics and the environment. As the new president of the National Committee for Ethics and the Environment has rightly said, 'That which is not scientific is not ethical'. Along with him, let us take this matter seriously.

From M. H, research assistant

Are scientists entitled to become involved with culture? As a young researcher (I am preparing my thesis in macromolecular chemistry) I was contacted by final year students from the Lycée Jacques Monod who wanted to do a PAE (active educational project) in their school, following the example of the INSERM youth clubs. In these, school students and researchers meet, and they fulfil very well, it would seem, their role of mediation between science as it is happening and science as it is taught. I was enthusiastic, and ready to commit myself to such an activity, aware that I would be fulfilling one of the tasks explicitly entrusted to researchers by the Research Orientation and Programming Act of 1982, namely 'to participate in the dissemination of scientific culture'. However, I was very quickly brought down to earth by my boss, who summoned me and severely admonished me for my intentions. There was no way I would be allowed to 'waste my own time, and the lab's, with people who in any case would understand nothing'. If I wanted a chance of getting the job of researcher with tenure – such places are now few and far between – I would have to 'concentrate all my efforts on my work at the laboratory bench, rather than degrade myself by doing something other than research'. 'You will have plenty of time later', was his parting comment (of course, he gives television interviews and public lectures).

How can the good intentions of the above-mentioned Orientation Act be implemented if researchers receive no training in the new tasks of cultural dissemination which they are supposed to carry out, and while no professional recognition is given to them for these activities? It is a shame that the reform of the Doctorate did not require that each candidate, in addition to writing a thesis, should provide evidence of having performed at least one public activity (writing a popular science article, giving a public lecture, participating in an exhibition etc.).

From M. I, civil servant in the Ministry of Research

I am very pleased that your publication has reported on the significant action undertaken by the public authorities in developing scientific culture, and would like to take this opportunity of thanking you warmly. May I however point out that state funding of initiatives undertaken in this field is not being reduced at all, as your editor seems to maintain. Since 1981 ministerial action has remained constant, dynamic and proactive. Our country is lucky to have establishments of the first order which, alas, were suffering from a chronic lack of finance and were falling into a moribund state. The collections of the Museum of Natural History were mouldering away under layers of dust. Now it puts together travelling exhibitions, and will soon open an important and innovative Gallery of Evolution. The National Museum of Technology was in need of roof repairs and few visitors trod the creaking floors of its great rooms. Now it is about to be completely renovated and will reopen its doors in 1994 after major refurbishment which, while preserving its atmosphere and the richness of its collections, will enable its exhibits to be set against a living history of technology. The Palace of Discovery, some of whose rooms still possess exhibits dating back to its hasty opening in 1936, has seen its interiors radically transformed and has embarked on a programme of restructuring. The 175 provincial museums of natural history and 200 museums of technology had, with a few exceptions, fallen into neglect. The most active of them benefitted from funding which secured their renovation. In the four years since its opening, the Cité des sciences et de l'industrie at la Villette has become the number one attraction worldwide and is drawing an increasing number of visitors. But above all, our Ministry's work has enabled programmes directed at youth to be supported throughout France (clubs, discovery classes, active educational projects). It has enabled the creation and development of Centres for Scientific, Technological and Industrial Culture – new establishments which are the only ones of their kind – which are a priority in our Ministry's programme, and whose enthusiasm is now infecting many regions of our country. Finally, I would like to add that the Ministry in charge of scientific research, while being in the firing line with this policy, is not alone. The Ministry of Culture also plays a considerable part. There is even an element that is peculiar to the politics of our country, which considers that culture, following the example of the Republic, is 'one and indivisible', and desires that science should not simply be added to it, but should be integrated into it. In conclusion, I would like to draw your attention to the following data appertaining to the year 1993.

Museums, Ecomuseums and museums of technology: about 1,200 establishments

Associations carrying out scientific activities: about 800 active

Centres for Scientific and Technological Culture: 19 in 15 regions (3 centres in 1983)

Financial aid from the Ministry of Research: 55 million Francs in 1992 (as opposed to 11 million Francs in 1983)

Educational projects for schools: 20 per cent devoted to scientific subjects (as opposed to 10 per cent in 1989)

Science exhibitions (designed and presented by young people in conjunction with CIRASTI) since 1985: 40,000 young exhibitors, 400,000 visitors to the 79 science exhibitions staged since 1985

Operation 'Fun with science' holidays (scientific tourism organized by the Family Holiday Villages association): 28,000 people (6,000 in 1990, 11,400 in 1992)

Operation 'Festival of Science' in 1993: 1.5 million participants, 1,500 activities, 5,500 researchers involved, over 2,000 articles in the national press

Authors

Bernard Maitte is professor of physics at the University of Lille and head of ALIAS, the Regional Centre for Scientific, Technological and Industrial Culture in Nord-Pas de Calais, at 75 Ch. de l'Hôtel de Ville, F-59650 Villeneuve d'Ascq, France. Jean-Marc Lévy-Leblond is professor of physics at the University of Nice and editor of the journal *Alliage*, at 78 route de Saint-Pierre de Féric, F-06000 Nice, France.

Science and culture in Germany: is there a case?

Renate Bader

'I like them tough to read', says my student, looking at me defiantly. Having just finished a long speech about how science writers could and should employ fictional devices in order to make their prose more lively and interesting, I am stunned. Does he really want to struggle through almost academic articles full of explanations and facts, I ask? 'Yes', comes the reply. Otherwise he would not feel the information was sound. Only after having understood a complicated article can he lean back and enjoy the feeling of having come closer to some of the best and most intelligent minds, he explains. It makes him feel less ignorant and better educated.

He is no exception. Whenever I am trying to introduce students to the idea of writing about science in an entertaining way, or am telling them they should portray science-in-the-making and scientists as normal human beings going about a daily routine, they grow restless and at times even resentful. To them the notion that science can be fun and can be represented as such seems completely alien. Science is something special, removed from the realm of ordinary life and concerns, and from their heritage and culture. It is perceived as either elitist or threatening. They want absolute truths from science, and worry about the impacts new technologies may have on society and nature. In doing so they mirror what a majority of the German public seems to think.

Colleagues from other European countries have repeatedly suggested that such attitudes should raise concern and initiate a flurry of activities aimed at improving the public understanding of science. They are surprised when they find it difficult to locate persons or institutions in Germany that have subscribed wholeheartedly to the cause of putting science into culture, in a country where the public is reportedly more opposed to science and technology than the public in many other countries.[1] They wonder whether things are indeed more narrowly defined by academic disciplines or social concerns in Germany, and want to know why. Answering these questions is not easy, partly because relevant research is mostly lacking, and partly because the issue is a highly complex one. But perhaps some preliminary observations and reflections will help explain why Germans tend to have either an exaggerated respect for or an exaggerated distrust of science, and may shed some light on the German situation which seems to be characterized by ambiguity: an academic approach to science communication coexists peacefully with a politically motivated environmental movement, which has infected major parts of society and threatens to develop into what has been termed an 'eco-dictatorship'.

Public perceptions of science

Science equals education

Traditionally, the public communication of science in Germany has been synonymous with public education in science, and has mostly been perceived as being highly intellectual and academic. Early attempts by scientists such as Alexander von Humboldt to popularize science through public lectures, and the initiation of local learned societies for public education in science such as the 'Urania' in Berlin,[2] were usually aimed at so-called 'educated circles', namely that well-to-do segment of society that had enjoyed a high-school or university education and could be expected to possess basic knowledge in a broad range of fields.[3] The same was true for popular science magazines and the science pages or supplements in daily newspapers, which were first published at about the same time.

When the first popular science magazine, *Umschau in Wissenschaft und Technik*, appeared in 1897 it promised readers 'a complete and reliable overview of advances in all scientific fields' written by chosen experts but in comprehensible language. The magazine saw its target group as upper and middle class citizens, who could be expected to be well educated and attentive to science. Right up until its demise in 1984, the magazine never looked for a broader, more general readership.

For the broader public a new form of adult education system developed. The 'Bildungsvereine' or educational societies first appeared in the mid-nineteenth century and offered classes in all the subjects that were usually taught at school, science and technology included. Here blue-collar workers and other underprivileged groups could for the first time supplement their education and develop their intellectual abilities. Later these societies were replaced by so-called 'Volkshochschulen' (public high schools), which now exist in all larger towns, and for some of their classes offer certificates towards a higher school degree or towards advanced professional training.[3] They are mostly taught by teachers from ordinary high schools or university graduate students, and thus have become part of the formal educational system, as have open universities and certain educational programmes broadcast on radio which regularly deal with science.

Clearly, science and education have strong historical links in Germany. Not surprisingly then, public lecturers, 'Volkshochschul' teachers and to some extent science journalists are recruited accordingly, namely on the basis of their formal academic science education. This certainly has consequences for any public understanding of science movement in Germany: rather then seeing themselves as serving a common goal, namely educating the public about science, the protagonists perceive themselves first of all as representing a specific academic discipline or professional group to whose standards and worldviews they adhere. The result is that 'science' is more narrowly defined according to academic disciplines or professional groups in Germany than it is elsewhere.

A similar statement can be made about a group of institutions which in other countries is increasingly discarding its aura of academia, and is experimenting with new approaches to public communication of science: museums. In Germany most museums are still clinging to their traditional scholarly role, doing research and caring for collections, while educational or communicative goals take second place. According to Willi Ziegler the director of the Senckenberg Natural History Museum in Frankfurt, his response to declining funds would be not to cut research but to leave exhibits unfinished instead. The human evolution exhibit at Senckenberg has, for example, been in progress for more than 20 years, since 1970.[4]

Learned societies organizing public lectures, traditional science magazines, 'Volkshochschulen', open universities, educational programmes on radio and TV, and science museums – all these still exist in Germany. But, although open to all, they only reach the attentive public and segments of the interested public,[5] since they all tend to be academic and educational. Science in Germany is authoritative, not fun.

Considering such traditions and circumstances it is probably less surprising that my student expected science stories to be written in a rather dry, boring style, so that he could enjoy the feeling of having learnt something new in an appropriate and not too easy way, and that in general science is not necessarily seen as part of its heritage and culture by the majority of the public. Instead many people, though in awe of science and the achievements of scientists, seem to be wary of the implications and consequences science can have for society and nature, precisely because they understand neither.

Science equals risk

The one science perceived completely differently in Germany is ecology. While science is seen as intellectual and academic, a strong environmental movement has caught the public's fancy. It is driven by political and moral concerns. Ecology is not necessarily seen as a science, but as a new holistic approach to all aspects of life and nature. It is precisely those who are most disenchanted with and critical of traditional research and its applications who are drawn towards the 'greens'. Science for them equals risk; ecology is the saviour.

Moreover, the 'green' movement has always been a grass roots movement. It really is wide open to all. Class and educational differences are irrelevant; being concerned about the state of the world is all that matters. Unlike traditional approaches to improving the public understanding of science, the emphasis is less on academia and education and more on what the public actually needs and wants. And ecology can be fun: in Germany we find eco-festivals, eco-theatre and eco-games.

Obviously the environmental movement is, at least in part, the reaction to the academic, authoritative and educational science with which the population has become increasingly disenchanted. Having been taught to trust science, scientists and scientific truths, the German public had to learn that German scientists invented dreadful weapons like nerve gas and the atomic bomb, and that science was instrumentalized and used during the Hitler regime.[6] Not surprisingly, science grew suspect. This attitude was reinforced during the sixties, when intellectuals started to question any kind of authority and opted for a preoccupation with social concerns. Disenchantment with science grew when Rachel Carson's book *Silent Spring* was published.[7] Soon after, the first politicians started to openly question conventional wisdom and eventually turned 'green',[8] and the Green Party was founded.[9] The safety of nuclear power was questioned and a debate about its risks ensued. Citizens' and advocacy groups appeared and fought against major technological projects, and a few years later there were an Eco-Institute and an Eco-Bank. Somehow ecological ideas had pervaded the whole of German society, and continue to do so today.

The ivory-tower mentality of scientists, which has resulted in a top-down approach to science education and public communication of science, is at least partly responsible for these developments. Instead of capturing public interest and imagination, scientists as well as science educators have mostly stuck to their academic disciplines and restricted their teachings to those already interested in the subject matter – or they have turned 'green' themselves.

Improving the public image of science

Science journalism

Needless to say, even the most hard-headed inhabitants of the academic ivory tower, as well as those in industry and in funding bodies for science and technology, have become increasingly worried in the face of such public attitudes and developments. But instead of changing their approach to science communication, or creating, in a concerted effort, a body like the UK's Committee on the Public Understanding of Science (COPUS) in order to foster new initiatives for reaching a broader public, they have turned to the mass media, and insisted that journalists should make a better job of explaining science. The solution was seen to be more science writers with better scientific training, and more space for articles and programmes about science and technology in the media.[10] Thus responsibility for the public communication of science was, to a large extent, delegated to one professional group and the media.

The Robert Bosch Foundation took up the mission, and started a programme aimed at improving science journalism in Germany. Mass communication researchers, established science journalists and public relations officers were invited to meetings to talk about the future prospects for science writing; fellowships for internships in the media were granted to young science graduates interested in science writing; and more experienced science writers received funds to join mid-career programmes in the USA. Other projects were also sponsored, and eventually the Foundation decided to help establish the first science journalism course at the Free University in Berlin, thus finishing its programme by ensuring that science journalism will continue to be taught in a university setting.

Meanwhile, science journalism in Germany seems to be well established. There are many young journalists with a high level of scientific education, a number of new editorial offices have been established, others have taken on more staff, and in a number of dailies the space for science articles has grown.[11] But while science journalists object to the notion that they should be educators,[12] many seem to adhere to academic disciplinary standards when writing about science. And since most are working for high-quality papers, magazines and broadcasts, again the higher educated, attentive public is better served than the vast majority of people, who also keep encountering science in their daily lives, and would like to understand it.[13] They still are not well catered for.

New initiatives

Worries about the state of public understanding of science have grown again recently within the scientific community and among science journalists. But remedies are mostly sought along traditional lines. The Max Planck Gesellschaft is now hosting the

European Initiative for Communicators of Science (EICOS) programme, which allows science writers to spend some time actually working in a laboratory, and the Forschungszentrum Jülich is planning a workshop which will introduce scientists to the work constraints and techniques of journalists. The science magazine *bild der wissenschaft*, in a similar attempt, has now opened its offices to scientists who are willing to do a media internship.[14]

Obviously all these activities aim at improving both the quality of science journalism according to scientific standards and the relationship between scientists and journalists. The people who will benefit most from this are that segment of the public that already is well served.

But there may be hope – and not only for the general public, but also for all those colleagues seeking contact in Germany with people or institutions wholeheartedly devoted to science communication. Science journalists coming back from the annual meeting of the American Association for the Advancement of Science,[15] the British Association's Science Festival and the Edinburgh Science Festival have started to spread the idea that Germany needs similar events. And scientists at their professional association meetings are now tentatively discussing possibilities for creating a German Association for the Advancement of Science.[16] Hopefully, a German COPUS devoted to experimenting with new approaches to the popularization of science will be established in the near future. At least, we at the Free University Berlin are working towards this end.[17]

References

1 Bauer, M., Durant, J., and Evans, G., 1991, European public perceptions of science: an international comparative study. Paper presented at the American Association for the Advancement of Science annual meeting, Washington; Cantley, M., 1991, The need for science in the media. *Science Communication in Europe* (London: Ciba Foundation), pp.9–16; Marlier, E., 1992, Eurobarometer 35.1: opinions of Europeans on biotechnology in 1991. *Biotechnology in Public: A Review of Recent Research*, edited by J. Durant (London: Science Museum), pp.52–108.

2 Buddensieg, T., Düwell, K., and Sembach, K.-J., 1987, *Wissenschaften in Berlin* (Berlin: TU Berlin).

3 Gruhn, W., 1979, *Wissenschaft und Technik in deutschen Massenmedien. Ein Vergleich zwischen der Bundesrepublik Deutschland und der DDR* (Erlangen: Deutsche Gesellschaft für zeitgeschichtliche Fragen e.V.).

4 Culotta, E., 1992, Worlds Apart: From Frankfurt to Honolulu. *Science*, **256**, 1269.

5 Miller, J.D., 1986, Reaching the attentive and interested publics for science. *Scientists and Journalists: Reporting Science as News*, edited by Sharon M. Friedman, Sharon Dunwoody and Carol L. Rogers (New York: Free Press), p.55–69.

6 Müller-Hill, B., 1984, *Tödliche Wissenschaft. Die Aussonderung von Juden, Zigeunern und Geisteskranken 1933-1945* (Reinbek: Rowohlt).

7 Carson, R., 1962, *Silent Spring* (Boston: Houghton Mifflin).

8 Gruhl, H., 1975, *Ein Planet wird geplündert. Die Schreckensbilanz unserer Politik* (Frankfurt/Main: Fischer).

9 Mettke, J.R., 1982, *Die Grünen. Regierungspartner von morgen?* (Hamburg: SPIEGEL).

10 Russ-Mohl, S., Wissenschaft und Öffentlichkeit–Zwischenbilanz eines schwierigen Dialogs, *Die Mitarbeit*, **32**(4), 341–353; Hömberg, W., 1989, Wissenschaftsjournalisten gesucht? Resonanz und Konsequenzen einer Untersuchung zum Stellenwert des Wissenschaftsjournalismus. *Unverständliche Wissenschaft. Probleme und Perspektiven der Wissenschaftspublizistik*, edited by Arno Bammé, Ernst Klotzmann and Hasso Reschenberg (München: Profil), p.217–234; Haller, M., 1987, Wie wissenschaftlich ist Wissenschaftsjournalismus. *Publizistik*, **32**, 305–319.

11   Hömberg, W., 1989, *Das verspätete Ressort. Die Situation des Wissenschaftsjournalimus* (Konstanz: Universitätsverlag); Fayard, P., 1992, The development of science reporting in European daily press. *Science and the Media – A European Comparison*, edited by K. Zerges and W. Becker (Berlin: sigma), p.97-104.

12   Korbmann, R., 1991, Science in magazines: words and pictures. *Science Communication in Europe* (London: Ciba Foundation), pp.41–44

13   Durant, J.R., Miller, J.D., Tchernia, J.-F., and van Deelen, W., 1991, Europeans, science and technology. Paper presented to the meeting of the American Association for the Advancement of Science, Washington DC.

14   Offene Türen im Elfenbeinturm. *bild der wissenschaft*, 1993, **1**, 22.

15   von Randow, G., 1992, Unverständliche Wissenschaft. *Die Zeit*, 5 June, p.52.

16   Schuh, H., 1992, Zerreissprobe für die Forschung. *Die Zeit*, 5 June, p.49; Editorial, 1992, Meinungskampf. *Frankfürter Allgemeine Zeitung*, 21 October, p.N1.

17   Göpfert, W., 1992, Öffentliche Wissenschaft, *dimensionen*, **2**, 9.

Author

Renate Bader trained in biology and journalism, and teaches science writing at the Institut für Publizistik, Malteserstrasse 74-100, D-1000 Berlin 46, Germany. She also works as a freelance science writer, and contributes to newspapers, magazines and radio.

Musings on the popularization of science in Spain

Jorge Wagensberg

I was born, along with many other Spaniards, in 1948: there were still twenty-seven years of boredom to go. At that time a lot of us believed that the cosmos was divided into two slightly disproportionate parts: Spain, and the rest of the universe. A curious division of space-time certainly, but time only seemed to move on abroad. Here history had come to a standstill. Daily life was horribly predictable in all aspects. Watches or calendars seemed to be highly sarcastic gadgets laughing at us while faithfully and accurately measuring a time that never changed.

It is worth pondering on the reasons behind that atmosphere of collective tedium and its consequences – trusting that we are not still living through it, that it will not come back, that it never went ... trusting that it does not still affect the greater part of the planet. I shall put forward an idea, knowing it may sound crazy at first, and aware that I have only a few pages in which to raise it to the category of being at least credible:

> One of the most tragic and pathetic roots of the human condition is embedded in the public's ignorance of scientific knowledge, and above all ignorance of the so-called scientific method.

I shall attempt to demonstrate this in three very short chapters.

Chapter 1

Knowledge is without doubt the clearest contribution people have made to biological evolution, and we owe our 'success' on the planet to the development of that function. What is knowledge? Let's take two initial strong working hypotheses as our starting point:

1. Reality exists
2. The mind exists (or, minds exist)

So knowledge is all that which is:

1. A mental representation (more or less accurate) of reality
2. Transferrable (more or less accurately) to other minds in non-genetic ways

The method used to build up knowledge defines the different kinds of knowledge. In theory then, there is an indefinite number of types of knowledge. Elsewhere I have tried to demonstrate that in fact there are only three outstanding kinds,[1] or to put it another way, any knowledge is the result of the thoughtful combination of using three, shall we say, pure methods:

1. Scientific knowledge (based on the requirement that three uneasy principles be fulfilled to the maximum degree possible: (a) objectivity, (b) intelligibility and (c) experimental dialectics). Scientific knowledge serves the purpose of forecasting a comet, making a horseshoe, etc.
2. Artistic knowledge (based on the one amazing principle that certain infinite complexities, not necessarily intelligible, can be transmitted through a finite representation, such as a musical score, a painting, a grimace).
3. Revealed knowledge (based on two effective principles: (a) a being exists who/ which owns the knowledge of all reality; and (b) this being sees fit to reveal to us (sometimes) (some) of his/its knowledge). Such is religion, rare inspiration, superstition... .

To roughly illustrate the point, let's say that so-called political knowledge is based on a scheme of 1–3, in other words in the scientific-divine area (autarchies being fascinated by the axis 3, and democracies searching for 1 as best they can; it is no coincidence that many dictators end up believing they have been sent by the divinity and that they have a divine mission amidst mortals; Franco liked to write 'By the grace of God' on seals and coins). We could also say that Picasso and Darwin used to hover around the 2–1 mark, that is in the vicinity of the artistic-scientific area, and that Van Gogh and Kafka were never far from the 2–3 artistic-divine scheme. Let's say the three methods have been consecrated by history and that all three, in an abstract way like this, are just and useful, but let's also say that once the objective of a certain particular knowledge is defined, some 'recipes' turn out to be highly suspect (over-divine public administrators), worrying (over-artistic bridge builders) or bewildering (over-scientific lovers).

Chapter 2

After thousands of millions of years of biological evolution, natural selection has consecrated a few extremely important basic functions, such as feeding, breathing and reproduction. How did evolution organize it so that the creatures of this world do not forget to eat and drink, to breathe, and to leave copies of themselves in time for the survival of the species? The question is by no means trivial. In fact, as everyone knows, matter, and living matter is no exception, is essentially lazy and always gravitates towards situations that require minimal energy. What has happened is that certain strong stimuli have been selected to guarantee these important functions: hunger and thirst for feeding, strong demand for defecation, micturition or breathing, and an equally efficient sexual stimulus for reproduction. Put in a different way, those creatures born without 'appetite' in one of these senses disappeared a long time ago. These stimuli, both pressing and not exempt from certain promises of enjoyment, seem to be an essential condition for survival in life. As for us, the human species, it has fallen to us to live through a real transhistoric event. We have perhaps not invented, but certainly consecrated, knowledge as a useful function for survival. Thanks to knowledge we have conquered the planet at a giddy speed: it is hardly a hundred thousand years since we acceded to it. Take note: a hundred thousand years of knowledge as opposed to hundreds of millions of years of breathing air and thousands of millions of years of sex and feeding. What has happened? Natural selection has still not had time to work on behalf of knowledge to consecrate it with some stimulus that is at once pressing and pleasurable. We have got to the crux of the question:

> At the level of the human species nothing has yet been formed which deserves to be called something like A THIRST FOR KNOWLEDGE!

And look where it's got us. It is a critical moment on a biological scale. We have become knowledge addicts and selection has still not given us a stimulus for consecrating it. (Look for example at the current programming of the Spanish television channels.) While that is the case, and if one day it should be, there is only one solution: to give ourselves one.

Chapter 3

I have exaggerated, it's true. You only have to look over the history of humanity briefly to see that, curiously enough, the most remote and independent of civilizations had beliefs about what lies beyond, gods, rites and all kinds of artistic objects and representations, and arms, implements and protoscientific practices. Let us then bring in the light of analysis of the first chapter. Yes, a certain powerful innate stimulus does exist for the revealed type of knowledge. How else would it be so impossible to find traces of a human group which does not manifest the worship of certain gods? There also exists some kind of stimulus, albeit weaker, for artistic knowledge. This allows us to get the exact measure of the issue:

> What really does not exist is a stimulus which favours the spontaneous practice of the scientific method.

There is nothing strange about this situation. Metabolic functions are thousands of millions of years old, the artistic and divine versions of knowledge only date back some hundreds of thousands of years, and scientific knowledge as we understand it today (the one which most affects us on an individual and collective level in our daily life) only goes back a few centuries!

And look where that's got us. A few individuals produce, at great speed, enormous amounts of scientific knowledge which entire humanity pays for, endures and enjoys. Such a contradiction gets bigger by the minute for two reasons. The first is obvious. How can citizens of a democratic society marked by science influence their future if their scientific training is not up to the norm of the Middle Ages? They need stimuli. The second is not so obvious. Modern man and woman, faced with any complex situation, usually resorts to methods of the artistic or divine kind, for which they do have some natural stimulus. Their options are clearly incomplete. Let's think for example about something as common, yet at the same time as complicated, as understanding our fellow humans (our family, other families, those from another neighbourhood, another town, another region, another country, another continent, another culture, another race). The history of humanity is a history of intolerance, and tolerance demands the application of scientific method to knowledge of our fellow humans, the only things that change truth when it is no longer compatible with observation of the world, the only ones whose duty it is not to respect its teachers, the only ones for whom everything could be changed. So scientific stimuli are necessary for two reasons, for the content of accumulated knowledge and for the method used.

Conclusion

I began these lines recalling the long night of dictatorship in Spain. At that time boredom and pathological ignorance were acute. A dictatorship, however light it may seem, seriously castrates the attainment of the two pillars of knowledge: making oneself a representation of the reality of the world with one's own mind and, if this can be managed, the possibility of transmitting this representation to other minds. Liberated from that system Spain woke up, and is with all speed catching up on lost time. Having overcome the acute pathology perhaps we have now reached the good 'Western' level of chronic pathology, because we now share the freedom to practise and know science with the same lack of stimuli shared by all the most advanced, industrialized societies. The media do not stimulate the creation of 'scientific opinion', perhaps because such a thing does not even emerge in the scientific community. Notice how the most sellable science is actually that which is the most divine or the most artistic.

Here lies the great conclusion: the popularization of science is a need that above all ought to be based on creating the stimuli that nature has so far denied us spontaneously. It would be a good criterion to try for those responsible for the media, which luckily, though perhaps mistakenly, still exist. It would be a good criterion for the science supplements of the daily newspapers, for science in general information, for television or radio programmes, audiovisuals, films, magazines with circulations that reach all levels, every kind of collectable series, study programmes in schools and universities (above all in the non-science faculties), libraries specializing in general science and above all non-academic centres devoted to diverse scientific themes. What is true is that these structures do now exist in Spain, as in many other countries in Europe; many of them are even pioneers on an international level; but also international is the fact that many of them have begun to go into decline without having even tried to transmit those stimuli for science. Science museums or centres, which in their modern form have a mission to stimulate laypeople, are few and far between and have difficulty resisting the temptation of 'show business'. Maybe it is these centres (many were conceived only for this) which should take the initiative which will carry the other media along with them, together with those who are obliged to be stimulating so as to be influential when people are choosing reading matter, activities or simply a television channel. One promising piece of information: the media which base what they offer on creating stimuli have been successful. And it suits people to have their minds stimulated.

My final thought is of an ideological nature. The inequalities between people are gigantic and it has always been thought that solidarity meant first providing food, and then perhaps, some day, providing knowledge, which may even include, some other day, science, as not all the world is yet ready for freedom and knowledge. This is probably an enormous mistake. Probably knowledge and freedom are not at all postponable nor conditional on something that has gone before. It is probably a good idea to take this principle as a starting point. In politics the Left preserve, I believe, a generic definition:

the grouping together of those who consider it a priority to achieve the greatest good for the greatest possible number of people. It is probably a good idea for the Left, somewhat confused and disoriented these days, to take it as theirs.

Reference

1   Wagensberg, J., 1985, *Ideas sobre la Complejidad del Mundo* (Ideas about the complexity of the world) (Barcelona: Tusquets).

Author

Jorge Wagensberg is professor of physics at the University of Barcelona, and director of the Museu de la Ciència de la Fundació 'La Caixa', Teodor Roviralta 55, 08022 Barcelona, Spain.

Portugal: past and future

Rui Trindade

For many years Portugal turned its back on Europe, and dozed in the shade of a ruralistic and anti-cosmopolitan dictatorship. Recently, however, it has redefined its own space within the concept of Europe.

Portugal's integration into the European Community has enabled quantum leaps forward, but these have also brought convulsions. Its financial fabric is fragile, the process of industrialization is faltering and the education system is inadequate. There are regional imbalances, and the citizens have limited means of influencing the decision-making process. But despite everything, the 1980s have been characterized by an accelerating pace of change. Economic growth, the development of road infrastructures, and technological reconversion and modernization in certain industrial sectors have promoted a powerful dynamic for transformation.

It could also be claimed that scientific activity has benefitted from this new-found dynamism. However, the injection of financial resources and the concentration of attention to the planning of research policies cannot conceal the tremendous gulf which separates Portugal from most of its European partners. There is still a long way to go. The scientific community is small, its links with the industrial sector are weak, and the shortage of international contact is severe.

In this context, it is hardly surprising that Portugal's scientific culture is rather limited, and we are only at the very beginning of our efforts to make science more accessible. But here, too, the 1980s have exerted an influence. Some initiatives have emerged as the result of a draft policy to encourage activities to make science more accessible: travelling scientific exhibitions, the publication of books and reviews, the setting-up of specialized video libraries, the encouragement of scientific journalism, and so on. JNICT, the national agency for the co-ordination and financing of scientific research, has played an important part. 'Science Week', a travelling exhibition which covers almost every scientific field, involved a considerable number of scientists in defining the various aspects of the exhibition. Preparatory discussions have considered the role of scientists in activities for making science more accessible. Although there is mostly support for regular participation in such activities, the organization of universities and laboratories, and major budgetary limitations, do not allow scientists to devote much time to popularizing science. The situation became more complicated in the early 1990s because universities and laboratories were placed under pressure to find resources in the commercial sector, so there is now even less time available for these predominantly voluntary, unpaid activities. Also, commercial organizations still have a very poor understanding of how to make science more accessible. Cultural patronage is given almost exclusively to artistic activities.

The most significant change has occurred in the communications media. In the press, more careful attention is paid to science and technology, and this has culminated in the inclusion in one of Lisbon's daily papers of a resident science column. There are now specialized science programmes on television, and better coverage of science and technology in the news. University museums of science, technology and natural history have joined international efforts (participation in the ECSITE network of science centres, and in a project to bring a dinosaur exhibition to Portugal, for example), and their successes very clearly demonstrate the public's readiness for and interest in scientific culture.

Two projects which are currently at the launch stage may exert a significant impact on all activities to make science more accessible. One is a mega-project: the completion of the Lisbon International Exhibition in 1998, with the theme of 'Oceans, Final Frontier of the Planet' to mark the five-hundredth anniversary of the discovery of the marine route to the Indies by the Portuguese navigator Vasco da Gama. This event will be on a global scale. Following Expo 92 at Seville, this project will entail reclaiming and converting a large area of Lisbon City near the Tagus. The fact that several exhibition sites for inclusion in this project will remain operative after the exhibition has closed could mean that Lisbon, by the turn of the century, might possess the widest range of museum facilities dealing with aspects of the ocean.

The other project is on a more limited scale, but it is so ambitious that it is bound to have considerable impact. A new 'discovery space' in an old factory on the bank of the Tagus, although it belongs to the pattern of modern interactive science museums, is intended to become an innovative space which is open to the new trends in museum activities. The Lisbon Municipal Authority (Camara Municipal de Lisboa – CML) has proposed the setting up of this 'discovery space' with the aim of making science more accessible, and it should open to the public at the same time as the Lisbon International Exhibition in 1998. The project demonstrates the Municipal Authority's interest in activities to make science more accessible and the priority accorded to the awakening of public opinion with regard to science and technology. In this context, CML's decision is in line with the international trend, particularly in Europe, for increasing commitment to supporting activities to make science more accessible. The clear purpose of this 'discovery space' is to join the trend of modern interactive museums.

Although the reference models for this type of initiative – San Francisco's Exploratorium, la Villette in Paris and so on – have promoted a definition of the space in which the museum proposes to operate, the work of preparation for the final project must include considerations beyond the functional conditions in which the 'discovery space' is set up. It is not just a matter of finding the most suitable strategy for founding a science museum at the beginning of the twenty-first century, but also of considering the very concept of a 'museum' as it has been understood hitherto, so as to determine whether the museum concept is still the best way to achieve our goals in the context of the runaway changes being experienced by modern societies.

Without wishing to pre-empt the considerations and discussions which have yet to start, I would nevertheless like to offer some ideas which might be a valid contribution to the agenda for discussion. Modern museums or 'science centres' originated and evolved from a pre-existing museum idea and model: a relatively stable, well-established configuration which defined a space with predetermined characteristics and clearly defined goals. This is the context in which museums have delineated their differences, established their particular features and defined their own independent strategies. But the concept of participation, the appeal of 'hands-on', and the drive towards interactivity have prompted the definition of a new way of relating to the public. But this model would imply more or less intentional incorporation of communication strategies into the museum project as such. And this has brought the new 'science centres' closer to other types of contemporary cultural facilities which are more dedicated to 'entertainment' than to active learning. This recent trend of evolution, combined with the need to provide continually updated information on progress in science and technology, has given rise to concept of the museum as a medium.

Finally, market pressures, with their associated laws of the survival of the fittest, have prompted both the more traditional museums and the new 'science centres' to define strategies for attracting and entertaining the public, and for venturing into the treacherous territory of cultural marketing. This fact rings an alarm bell about one of the risks which beset science museums: the risk of creating 'Disneylands'.[1] And this is another important offshoot in museum development.

All of this prompts us to consider the future for museums – for all types of museums – and their operation in co-ordination with their local community. Without going too far into the tricky business of trying to forecast the future, we might assume that one of the predominant features of the future will be globalization, and the extension and intensification of technological performance in the field of information and communications. The influence to be exerted by such factors in the redefinition of conventional boundaries and concepts, especially in the blurring of the concept of public space and private space, will tend to increase. It is also possible that the increasing availability of technology will dramatically accelerate two trends which are only apparently in conflict with each other: a trend for nomadism based on the miniaturization and portability of equipment,[2] and 'polar inertia' which corresponds to an increase in connections, networks and virtual technologies, encouraging geographical stasis and removing the need for travel.[3]

The relationship of the individual with public space will tend to increase in status and acquire a fresh dimension. Access can be made either from inside (i.e. from home), or from the outside, via a satellite located far from our planet. The act of going to a museum in the physical sense may therefore lose its unique character and become one of a range of options. Electronically organized 'virtual visits' may become more and more frequent.[4] If we look at what has already happened in many museums, even in the field of the arts, with the transfer of complete collections on to digital storage media, and if we consider the fact that such collections will shortly be accessible by telephone, we can immediately see the direction in which the relationship with museums is bound to go.[5]

These experiments, which are transforming museums into slides which can be viewed from a distance, are a clear indication that the future lies with interactivity and user friendliness, and that museums are going to become active platforms for an intense social dialogue.[6] The channels which have been opened up by the electronics revolution will make it possible to devise not only spaces for consultation of an unrivalled capacity (using CDROM, CDI, etc.), but also spaces for discussion and intervention (electronic mail, 'bulletin boards', interactive TV, etc.). The museum will be wherever the individual is.

Museums are spaces which make available information, which stimulate interactivity or which provide advanced learning. It is likely that museums will evolve in the direction of producing information themselves. The increasing link between science and technology and the modern world, and the distribution of technologies, have weakened the role of conventional sources and will put an ever-increasing pressure on museums to function as specialized reference centres. This will give us a museum which is networked to other museums, universities, research centres, media, municipal resources and so on. This opens up the potential for a completely new type of museum. In the future, it may become the creative source of events and the leader of discussions that take place away from the physical museum – possibly anywhere, even in the home.

In essence, the museum must now be thought of a 'medium', not only in the strict sense of its internal communications strategies, but also as a source of information, as a centre of events and as a distributor of ideas. Science museums also have the option of developing into cultural producers, for example by adding to operative space those facilities which are at present peripheral experiments in transfer between technological experiment and creative expression.[7]

By extrapolation, and irrespective of the components which currently go to make up a science museum, we can conceive that in the future they will be regarded as multimedia cultural resources. Their information will no longer be exclusively available on site, but will also be available for users to call up from home, in schools, in media centres, etc. The museum will no longer be purely a reference centre, but also one for dialogue which stimulates creativity and enables fresh forms of training to be practised: training tailored to individual needs, and an exploratory form of learning.[8] At a time when scientific knowledge is increasingly bound up with political and social choices, the museum itself will have to exert the function of a producer of information, of cultural intervention and of creative stimulation.

For a country such as Portugal, the creation of such a 'discovery space' meets the absolute need for cultural modernization in order to make up for lost time. But the concrete incarnation of this project in its most radical dimension also presupposes the development of a real international network, especially for Europe, capable of making the various resources commonly available. From this viewpoint, the creation of a real electronic network between museums on the European scale appears to be essential in order to enable European scientific culture to become a more tangible reality.

In a recent debate, an interviewee was explaining that Portugal's incorporation into the European Community had enabled greater structuring and rationalization of Portugal's economy. Perhaps it should have been added that in order to provide consistency and a different quality of activities to make science more accessible, it is also necessary to achieve more effective integration of scientific culture in Europe.

References

1   Miles, R., 1987, Museums and the communication of science. *Communicating Science to the Public*, edited by the Ciba Foundation (Chichester: Wiley).

2   Attali, J., 1990, *Lignes d'horizon* (Lines on the Horizon) (Paris: Fayard).

3   Virilio, P., 1990, *L'inertie polaire* (Polar Inertia) (Paris: Christian Bourgois).

4   Rheingold, H., 1991, *Virtual Reality* (New York: Secker & Warburg). The most recent developments in the field of virtual reality, with the emergence of concepts such as 'televirtuality', have opened up one of the most valuable fields to be utilized in the context of museums.

5   Editorial, 1993, Art goes high tech. *Art News*, February.

6   Sterling, B., 1992, *The Hacker Crackdown* (New York: Bantam). The emergence of cyberspace as a new zone for in-depth exchange, for transfer of information and user friendliness via electronic networks, makes it possible to utilize uncharted dimensions in the global village.

7   See, for example, the operation by the Americans of the 'Electronic Café', and the interactive creative potential of new technologies.

8   Authier, M., and Levy, P., 1992, *Les arbres de connaissance* (The Trees of Knowledge) (Paris: La Découverte).

Author

Rui Trindade is a scientific journalist, director of the TV programmes 'Aventuras do Conhecimento' and 'Ambientes', which are broadcast on the Portuguese Channel 2, and a member of the working group for creation of Lisbon's new 'discovery space'. He is based at Fabrica de Imagens, Largo dos Lóios 4, P-1100 Lisbon, Portugal.

Science and culture: one view of Italy

Paolo Galluzzi

Everyone knows that Italy is the land of museums *par excellence*, yet so often when the thousands of Italian museological institutions are cited as proof of the excellence of our country internationally, science museums and scientific, technological and industrial collections are generally not taken into consideration. This omission is the consequence of a number of factors, the first of which is a cultural tradition which has gradually excluded technical and scientific history and activities from the landscape of historical research. The historiography of science and technology has emerged only very recently in Italy, although it can pride itself on a number of illustrious pioneers who are, almost without exception – and not by chance – an expression of the culture of positivism, from the nineteenth and twentieth centuries.

Thus, the history of science has only been taught in Italian universities since 1980, although it has been an established subject in other countries for decades. Traditional historiography has centred on ethical and political questions, and has neglected the history of science and technology. Today too, Italian history pays only marginal attention to the history of research practice, of scientific institutions, and of the public financing of scientific and technological research. Clearly lacking are the studies, data, directories of sources, biographical dictionaries and bibliographies which would help us to understand the development of these factors which over time have drawn the contours of contemporary science, with all its characteristics and problems. The traditional hostility in our country towards science (it has never established firm and durable roots in Italy) has resulted in the marginalization of the study of the problems of scientific research and its technological applications.

So it is not difficult to understand why we see in Italy such a marked contrast between the immense wealth of documents and collections which are preserved and the shortage of science museums or institutions and centres for the diffusion of scientific culture. The institutional weakness of science museums is the outcome of a constant absence, since Italian unification, of any policy to protect and promote the heritage of the history of science. In this area Italy has done almost nothing, since unification, to keep up with England, France, Germany or the United States, and the fortunes of the science museums which had sprung up in Milan, Bologna and Florence in the pre-unification period experienced a sudden downturn after unification.

Despite appeals voiced by certain particularly aware Italian intellectuals (Casati, Metteucci and, later, Mieli and Garbasso), there was a dearth of proper science museums in Italy through to the end of the 1920s. At around this time a significant trend emerged, for fascism put in place over two decades a whole series of initiatives which placed great value on scientific heritage of historical interest, and on the diffusion of the culture of science and technology. This gave rise to almost all the organizations which still operate, with varying levels of effectiveness, in our country today. These efforts were not disinterested. The fascist regime wished to demonstrate the benefits that

Italian culture as a whole had drawn from political guidance – guidance which was, in short, virile and authoritative. It also recognized a propaganda tool in the idea – an idea which had been dear to the Risorgimento – of Italian pre-eminence not only in civilian life but also in science. This pre-eminence was the cue for action, and a series of institutions was created and inaugurated with great pomp. This was the motivation behind the establishment of the institutions which still conserve and exhibit the scientific heritage of our country: the National Science History Institute and Museum in Florence, founded in 1930, the end result of Italy's first exhibition of the history of science; the National Museum of Science and Technology in Milan, built in the late thirties at the initiative of Guidi Ucelli di Nemi, and prompted by the major exhibition of the regime, 'Leonardo and Autarchy', organized in 1939; the Domus Galileana, founded in 1942 by Bottai and placed under the presidency of Giovanni Gentile; and the Tempio Voltiano, founded in 1927 – the word *tempio*, which means temple, speaks volumes about the purposes of this initiative.

I could also mention E 42, the universal exhibition of Rome, which was meticulously planned, but which could not be held because of the war and the fall of the regime. The plans for the Science Exhibition in E 42 (which, according to its organizers, was then to become a permanent museum) are held in the archives, and indicate that the intention was to erect a solemn monument to Italian scientific genius, showing the merits of a regime which, after centuries of intolerable usurpations by foreign countries, had finally redeemed the 'Italian pre-eminence' which this proud and public exhibition would celebrate.

The problem which must not be overlooked if we are to understand the present is that nothing changed after the war. The new democratic and republican state basically left this institutional legacy and its motivations intact. It did not even consider reforming its statutes. Naturally, in the more thriving institutions, transformation came from within, thanks to the intellectual forces at work there. This was the history of science with critical maturity, the development of its own dignity as an independent discipline and its emancipation from any type of manipulation, propaganda or chauvinism. All this provided a fundamental contribution to the cultural rebirth and restructuring of science museums in Italy. The state authorities, however, persisted in avoiding any real renovation of these establishments: they did not do what was necessary to transform the 'temples' (as they were originally conceived) into 'workshops' producing serious, critical and up-to-date scientific and history of science information for the general public.

While a 'temple' can normally be closed or opened by the authorities for certain solemn occasions, workshops require rigorous looking after, constant care and on-going exchanges between the public and the world of research. State support for a number – and, incidentally, a very small number – of institutions has remained, in its nature and in its spirit, largely similar to the support given during the twenty-year period of fascism. Furthermore, it has remained constantly inadequate, and at the same time uncertain and intermittent. These institutions have received absurdly small sums to cover their running costs, while sometimes quite considerable resources have been handed out to them for ceremonial displays (centenaries, exhibitions for tourists and political, commercial or even sporting events). These financial conditions, with these aberrant features, betray the survival of a distorted view and low opinion of these institutions. Yet they ought to be playing the important role – with numerous and significant institutional incentives – of fulfilling their duty of protecting the enormous and neglected scientific heritage in Italy, and of responding to the need for a wider and critically motivated diffusion of the culture of science and technology of today.

This constant neglect is illustrated by the story of the creation in 1975 of the Ministry for Cultural Heritage and the Environment, when it was separated from the Ministry for Public Education. No provision was made for the identification of the institutions, peripheral structures and central bodies which would be responsible for the delicate matter of conserving and promoting the scientific cultural heritage. Indeed, even today, no reference is made in the legislative documents of the Ministry for Cultural Heritage and the Environment to this area of history (with the exception of the recent Law 84, in which scientific cultural heritage is explicitly mentioned). This ministry, moreover, does not envisage implementing any particular framework to meet the specific needs of science museology. No provision is made in its expenditure budget and it has no management boards or committees in this area. Finally, not one expert on science museums is on the National Cultural Heritage Council.

This scenario, which is depressing from the cultural and institutional perspectives, has been changing gradually since 1988 thanks to the spirited and determined commitment of Antonio Ruberti, who became Minister for Scientific and Technological Research. Minister Ruberti, who had already undertaken major initiatives to maximize the scientific cultural heritage and to promote the culture of science and technology for over ten years as Vice-Chancellor of the University of Rome 'la Sapienza', set himself as a strategic objective the implementation of a promotional project throughout this field.

The difficulties and the obstacles to be overcome were quite considerable: the latest surveys revealed that the total number of museums in our country stands at over one thousand. None of these organizations is directly dependent on the Ministry for Cultural Heritage, which exercises a supervisory function over only an extremely limited number of institutions. University science museums fall into the state category; they constitute a fundamental core of institutions which otherwise generally have no autonomy, as they are under the control of the university institutes and departments. Other 'potential museums' run by the state include: those in educational establishments of all kinds and all levels, which come under the Ministry for Public Education; major museums and remarkable collections of scientific cultural items relating to engineering, aeronautics, geography and so on, which come under military administration; and the museums which come under the administration of the Ministry for Health. A considerable number of science museums, especially natural history museums, belong to local authorities which are also involved, with the Regions, in numerous Foundations throughout the country, and which are looking to promote our scientific cultural heritage and to make it productive.

To prevent the institutional fragmentation of collections and to promote synergy among the various administrations involved, Minister Ruberti instituted, in Spring 1988, a National Committee for the protection and promotion of historic scientific heritage and for the diffusion of scientific culture, which brings together those managers of the main public and private, national and local Italian science museums, and historians of culture, science and technology who are interested in having their documents and material evidence put to good use; it also includes illustrious academics engaged in the battle to guarantee capillary diffusion of a quality scientific culture, experts in scientific communication and representatives of the main national research bodies.

The task entrusted to the National Committee was to define the state's strategies and priorities for gradually implementing a national system of science history museums and centres, to be responsible for protecting and promoting the vast science history heritage of our country, and at the same time to help to create permanent institutions for science communication training and scientific cultural diffusion. In Autumn 1988, the National Committee presented its first document to the Minister. It indicated the procedures necessary for the systematic promotion of this sector of cultural activity which had traditionally been neglected in Italy. It recommended, in particular, two immediate lines of action: firstly, the renovation, by means of financial and legislative intervention, of the few scientific museums worthy of this title; and secondly, the promotion of the birth and development of museums modelled on the museums which function as centres for the communication of scientific information such as are found in many countries in Europe and North America. The report drew the Minister's attention to this in particular. The Committee underlined the lack of these institutions in Italy, which could have played a decisive role in the establishment of a policy on scientific information and the diffusion of scientific and technical culture. In the document, the National Committee recommended that an 'outline law' be drafted with a view to overcoming or at least reducing the difficulties stemming from the multiplicity of institutional responsibilities for science museums (various ministries, regions, local organizations, universities and private individuals). It also drew attention to the delicate matters of identification, cataloguing, and restoration of scientific instruments and other material evidence. Finally, a feature it highlighted was that training should be provided for an initial core of specialists in the protection and promotion of the technical and scientific heritage and the methodology of information and of the diffusion of the scientific and technical culture, with the appropriate integration of information technology. Our country lacks in-depth experience and a wide skill base in these areas.

An initial operational response was the launch in 1990 of a strategic project by the National Research Council (Committee 15) entitled 'Scientific Culture and Scientific Cultural Heritage'. The NRC project provided considerable financial support for some 20 active units (in total, over 400 practitioners and researchers) working throughout Italy in two directions, one research and one practical: the surveying, cataloguing and use of the science history heritage; and the development of resources and instruments for the communication and diffusion of the scientific culture.

The results achieved in the space of two years of intense activity were pleasing. A map of the main depositories of documents and material evidence was drawn up. A large number of collections were catalogued and restored. Catalogues of the major collections were published (thus reducing the risk that they would be fragmented) and at the same time a large number of multimedia aids were prepared to illustrate effectively, for the public, the cultural importance of the collections. The research effort concentrated above all on scientific instruments, bio-medical and natural history items, and technological and industrial collections. In addition, the units working on methodologies of science communication developed and tested innovative models, considered more up-to-date designs for museums of science and technology in a series of debates (with huge international participation), and developed experience in the role of the media in strategies of scientific cultural diffusion.

During this time, the 'strategic project' helped draw attention to the history of research institutions, to the problem of the protection and promotion of scientific and technical archives, and to the need to develop historical research on the development of the research system, with its international links, in Italy after the unification.

The campaign to which the 'strategic project' gave rise will be followed by a research commitment over a number of years, the feasibility of which has already been defined. It is a project finalized by the NRC (lasting five to ten years) which plans to develop the investigations and experiments required to obtain the skills necessary for the establishment of a national system of science history museums and centres, and to train those who are to run them. In the 'finalized project', attention is also focused on research into the problems of the modern history of science and technology (history of institutions, laboratories, education, financing, researchers and operators in the field of science and technology).

The first effective research tool, the 'strategic project', was quickly joined – also at the initiative of Minister Ruberti – by a legislative instrument in the form of Law 113 of 1991, governing the diffusion of scientific culture. The law is intended to support structural operating interventions of the system put in place, favouring in particular collaboration between the various 'focal points'. This support, in the first two years, was crucial in establishing an awareness of the worth of existing institutions (the National Science History Institute and Museum in Florence, the National Museum of Science and Technology in Milan, and the major university museums in Pavia, Bologna, Florence, etc.), and for the development of scientific information and documentation centre projects. To mention just a few particularly significant examples: the new institutions in Naples (IDIS Foundation), Trieste (Laboratory of Scientific Imagination) and Rome (MUSIS) are currently moving from the stage of abstract project to that of structural implementation, and are organizing, as in Naples, activities which are well received by the general public.

The policy which I have outlined was accompanied by a series of promotional and mobilization campaigns, and special mention should be made of Scientific Culture Week. Started in 1991 by the Ministry for Universities and Scientific and Technological Research, the *Settimana* was an outstanding success. In 1992 there were over 600 events, and in 1993 the total exceeded 700.

The rapid and unexpected explosion of interest in scientific and technical culture poses some difficult problems. The first aspect to be instilled, above all on the cultural level, is the dynamics of reappraisal of the technical and scientific tradition and the development of the diffusion of scientific and technical culture. We find ourselves looking at a characteristic feature of the Italian strategy in this sector: a successful campaign of basic scientific education, which does not only look at science and technology as they stand today, but as long-term processes in which the most recent results are regarded as the final outcome of a process of development, the whole of which is viewed with equal interest. The act of maintaining a link between the present and the past of technical and scientific research constitutes, *inter alia*, one of the best guarantees of diffusion of critical and insightful information – information which avoids the pitfalls of apologia or of putting science 'on trial'.

In a second phase, after defining the content of the centres of the national system we wish to establish, and having located the fundamental focal points, the structures need to be put in place (centres, personnel, services) so that the institutions can operate and forge links. Otherwise, the effort invested – however important it may be – will be wasted. In the fragile economic situation in which our country finds itself today, it is vital that we select a small number of institutions (some already in existence, others yet to be founded) to be interconnected as service centres. They will be linked to a multitude of centres throughout the country. The Italian focal points must of course be open to the main European or international focal points, in the context of a cultural integration which precedes political and economic integration. Indeed, scientific culture can help to accelerate this process. The focal points are to guarantee the preparation and the circulation of exhibitions, links with the media and the provision of services, such as the training of personnel.

The primary focal points of the system will also be concerned with optimizing collaboration with the main international institutions, which has hitherto been rather haphazard and over-dependent on personal contacts. National and international exchanges will mean a widening of the range of mechanisms for diffusing scientific and technical culture, with clear benefits to the user, and a tangible reduction in the costs of running the system.

Together, the campaigns implemented in Italy since 1988 have already produced extraordinary results, and opened up promising prospects. It now seems possible to make up the ground which Italy has been losing steadily to many other countries over the past few decades. Once the phase of mobilization and definition of orientations and objectives is completed, the effort which has traditionally been the most difficult one for our country has to be made: the transition from an intermittent policy of support, however generous and committed it may be, to the definition of permanent financing to create institutions able to develop long-term quality activity. The outcome of this challenge will be the decisive factor in enabling the results achieved so far to be of long-term benefit. And let us also hope that the authorities concerned do not forget the civil importance, alongside the cultural importance, of the promotion of scientific tradition and the diffusion of scientific culture. These objectives are of particular importance, and will have a strategic role to play in the development of an integrated Europe.

Author

Paolo Galluzzi is Coordinator of the National Committee for the study, protection and diffusion of science and science history culture, Professor at the University of Sienna, and Director of the Florence Institute and Museum for the History of Science, Piazza dei Giudici 1, I-50122 Florence, Italy.

The cultural challenge of science: the Italian case

Franco Prattico

The film is called *Death of a Neapolitan Mathematician*, and it has just been released in Italy (and will probably soon arrive in France). It is the story of the last week in the life of an Italian mathematician, Renato Caccioppoli, who committed suicide in Naples in 1959. The film was made by a young director, Mario Martone, and won awards at the Venice Festival.

There is no lack of the ingredients that go to make up a successful work, from the point of view of the public as much as the critics, and the scriptwriter and director have astutely mixed these ingredients to bring to life on screen one of those strange and enigmatic personalities (or at least that is how the general public regards them): a mathematician. And a certain astuteness is needed to film a personality such as Caccioppoli, who, during the 1950s, was one of the protagonists of the intellectual and political renaissance which took place in Naples in the wake of the war and fascism, but who was also better known in Italy and in his home town for his eccentricities rather than for his scientific work. Nonetheless, Caccioppoli had been in the top rank of mathematicians. He had produced pioneering work in practically all fields of mathematical analysis (topological, real, complex), from spaces of infinite size to calculation of variations, integration theory and geometrical measurement, often making progress that put him ahead of his time. And he was not 'only' a mathematician: he was a versatile personality, a gifted artist, a refined musician, and a cultured European. He was a strict anti-fascist: he had been arrested and, thanks to pressure from his powerful family, locked up only in a mental hospital rather than in prison. This episode helped to fuel the legend. Whilst not an 'orthodox' Marxist, during the immediate post-War period he became the reference point and the leader of a group of young Neapolitan communist intellectuals to whom he was able to impart not only modern sciences but also Mallarmé and Thomas Mann, nineteenth-century English poets, Debussy and Brahms.

For Caccioppoli there was no boundary between art and science, between the intellectual world and political and human commitment. He was a warm, unpredictable, ironic personality. He would travel around Naples wrapped up in the same faded overcoat, but was nonetheless indefinably elegant, almost a dandy, cigarette permanently in his mouth, and a feverish glint in his eyes. He left the university at which he had been teaching analysis to take up a professorship in literature and art, politics, epistemology and life, surrounded by bottles of cheap wine in the back street bars of Naples, to instruct his group of young friends which was made up of not only mathematics and physics students but also poets, painters and musicians, together with workers and sailors. Altogether, he was one of the essential creators of an Italy which was emerging from the war and from fascism, in a retarded and disoriented condition, but avid for European novelty and culture. And although communism was regarded with suspicion by blinkered bureaucrats, the view taken by Caccioppoli and his young friends was that

71

the 1950s represented the opportunity to redeem a nation which had been humiliated by fascism and by a narrow provincial culture.

The film dedicated to Caccioppoli is no doubt a success in terms of cinema. However, it does not examine the reasons for his death, for his political and human disappointment, nor the crisis in his scientific work (which was probably the real cause for his suicide), while the latter part of his life was undermined by his drinking. The character portrayed in the film could be a professor of romance philology or a failed poet, a journalist or an accountant abandoned by his wife, rather than a mathematician. The 'mathematician' of the title is a mere label propped up by a few scenes filmed in Naples University lecture rooms.

And why is reference made to mathematical analysis, scientific work or the turmoil of research in the context of a film or even in a novel or a work of music? What can be the common ground between differential and integral calculus and emotions, sentiment, and all the colour of life? And *Death of a Neapolitan Mathematician* resurrects the problem of the relationship between science and art, and their apparently incompatible levels of activity.

Do we discern the signs of impossibility? In fact, the adventure of science and scientists has been covered only by minor educational films which are screened around the colleges and yawned at by students. The world of emotions appears to be excluded from the frigid theatre of research. What film producer, for example, would dare to release a film about Boltzmann, about his conflicting relationship with the scientific community of his age, including, as the protagonist, his intellectual and human turmoil, and the dazzling importance of the redefinition of thermodynamics? Is it possible to tell the story of the birth of music for Mozart but not that of general relativity for Einstein?

In fact, the reason it is impossible to tell a story of science and the intellectual effort which bolsters it, and impossible to interpret as art the resulting worldview, is probably due to the way in which science is lived both by the scientific world and by the community of the 'humanist' culture. For scientists, the combination of a jealous defence of their specialization and a fear of the inevitable approximations in the interpretation into images, words or emotions of the hard and fast content of their work, must culminate in a distrust of those who would attempt the task, and such distrust has often been justified. From the point of view of the humanist intellectual or of an artist, science is a separate world with magical or even sinister characteristics: locked up in a code which cannot be deciphered on the plane of expression, it resists the application of everyday experience and sentiments which appear to be the preserve of art. (In Italy, an exception could be found in contemporary music amongst the disciples of Nono and Bussotti, inasmuch as they use stochastic methods and have to be conversant with acoustic physics.)

But the separation of the two cultures in Italy is particularly accentuated by the combination of two factors: the persistence in the training of intellectuals and in schools of the strong influence of Benedetto Croce – of a retarded idealism which relegates scientific work to the status of something incapable of achieving truth and beauty, imprisoned by a quantitative method which chops off the richness of the world. Such an attitude, which is often not explicitly stated, is powerfully influenced by the structural weakness and ineffectiveness of the Italian school system (though with some notable exceptions, especially in some universities such as the College of Pisa, which is now directed – and not by coincidence – by a physicist, Emilio Picasso, who was the director of the LEP Project at Geneva). This ineffectiveness is particularly dramatically visible at early levels when the very languages of science are rendered inaccessible, and a divergence is set up and becomes more and more pronounced, until by the university stage the two universes cannot communicate with each other (with the possible exception of some linguistics courses). A student who has opted to study science, whether by vocation or by necessity, is locked in his own discipline, perpetuating a type of arrogance in specialist knowledge which has no point of contact with culture in general: the result is the production of possibly the world's most ignorant engineers and technicians.

Furthermore, generalized representations portray the engineer as having a useful part to play, which brings a certain degree of respect, while the 'pure' scientist is regarded as a being who is endowed with secret abilities to decipher nature, although they are essentially mad or monomaniacal. Furthermore the scientific community in Italy as a whole makes virtually no serious efforts towards emerging from the academic shell, and is reluctant to explain its work or its goals, and can only be brought to do so when the release of public money depends upon the provision of further information in the scientist's special subject. Otherwise (and this is not an exclusively Italian failing), researchers are often isolated from the significance of their research and have a tendency to forget the global direction of scientific work, and to ignore cognitive processes altogether. So science comes across as esoteric and highly effective, but comprehensible only to the initiated.

However, this image of science as 'sorcery', which can find solutions to all ills (provided that enough public money is poured into it), while it amuses the scientific community, is actually sustained to a certain degree by the media, and is mainly upheld in the world of politics. Italian politicians are generally ignorant of the language of science (and often of that of any form of culture). They regard scientific culture (and indeed culture in general) as a pleasant ornament, but one which is not worthy of any real interest: it has nothing to say, and is boring and obscure. Few political Ministers for Universities and Research have emerged from the scientific world: one was Antonio Ruberti, but he was soon replaced by a journalist whose scientific publications amounted to a few small works on Catholic syndicalism in his own area; another is Umberto Colombo who took office in 1993. It is easy to understand that this absence of expertise influences not only the organization of research, but also the separation between scientific and literary culture, and the nature of the respective points of view.

Fortunately, this grim scenario is redeemed by a part of the Italian intellectual world (philosophers, semiologists and epistemologists) which has a sound foundation of knowledge of the horizons and problems of contemporary science, and participates actively in international debate. But naturally, in common with all philosophers, semiologists and epistemologists the world over, this activity is more or less limited to the cloisters of the academic world. There are some small but culturally influential groups who broach the topic of synthesis between the two cultures, in terms of the relationship between scientific knowledge, the image of the world and its representation. For example, in the context of Spoleto's Festival of the Two Worlds 'Spoleto Scienza', over the last three years a programme of conferences and seminars open to the public has achieved remarkable success in prompting scientists, literary figures, artists, writers and philosophers to participate in debates which include the public, who often take a very active part. Such 'encounter' seminar debates are now starting to be organized by private or public sector foundations, especially in the north of Italy, and seem to be primarily of interest to young people. Some publishers (Einaudi, Adelphi, Laterza) have for some years now made it their policy to publish literature which makes science generally accessible and have engaged in experimental programmes, though with a slight inclination to opt for English-speaking culture; conversely, a traditionally scientific publishing house (Boringhieri) has produced some experimental literary anthologies. Over the past few years, too, there have been occasional articles, interviews and short essays on the problems of scientific research in daily national newspapers and some weekly periodicals. There are journalistic efforts under way to make science more accessible, with a trend away from indulging in journalese when describing the latest discovery (of the sort which has occasionally produced comical misunderstandings such as when journalists raved over 'cold fusion' or 'water memory'), and a pattern is emerging of a more concentrated and even more critical type of journalism concerning the processes involved in 'practising science', 'science in practice', etc. Some cultural reviews such as *Prometeo* or *Sfera*, are now aptly characterized by efforts to show that scientific work constitutes a necessary culture.

One of the main centres in which scientific research is reinterpreted as culture is to be found at Trieste. The Laboratory of Scientific Imagination, founded and directed by physicist Paolo Bundinich of the International Centre for Theoretical Physics, also publishes a periodical which is released monthly as a supplement to a local daily paper. But the most interesting initiative in this direction is that taken by the interdisciplinary laboratory of SISSA (the International College of Advanced Scientific Studies) which is led by Claudio Magris, and has become a focal point for cultural encounter and productivity between scientists, literary specialists and artists. This laboratory is going to give rise to a College of Scientific Communication and Journalism, with the ambitious goal of training a new generation of intellectuals, with a humanist education and a scientific culture, who are capable of relating the two worlds to each other and of contributing to the joint stock of culture the hypotheses, problems and results of scientific research, taking them out of the context of their specialization and imparting to them a value in terms of the exploration of the world. This is an ambitious project, and certainly one which is hard to achieve, especially because of the reluctance of the media and the universities to allow conceptual revolutions (which are the real innovative wealth of science and its contribution to the culture of our age) to emerge from the world of the 'experts' and to be brought into open discussion in society at large (and also made accessible to poetic expression). A challenge by the arts and by humanist culture to the worldview which arises from scientific work does not seem to be forthcoming; this is due to ignorance and timidity with regard to scientific knowledge, or because the images produced by scientific work reawaken the conflict with the 'naïve' perception applied in our everyday interpretation of reality. Whilst psycho-analysis has deeply infiltrated literature and even visual arts, we find that physics and biology – although responsible for an even deeper transformation of the image of world, humankind and life – are pushed to one side, to be used but not 'believed in'. However, this stunning revolution – as Italo Calvino commented before his death – the relativity and space-time continuum of such a revolution and the invasion by the discrete and probabilistic science of quantum theory, and the immunological birth of the 'self', may bring their message to the very heart of artistic creation. So far, however, the main conceptual constructions of modern science appear destined to remain nothing more than laboratory instruments. Thus it is forgotten that science, in common with art and poetry, is essentially the process of research into the architecture of the world and the meaning of humankind.

Author

Franco Prattico is science editor of *La Repubblica*, 11/b Piazza Indipendenza, Rome, Italy. His latest book, *Dal caos alla coscienza* (From Chaos to Consciousness), was published in Italy and Germany.

Characteristics of public understanding in a changing society: an inquiry in East and West Berlin

Karlheinz Lüdtke and Renate Müller

A few years ago sociologists of science in the former GDR started to take notice of a subject in which Western opinion research had been interested for more than four decades: laypeople's understanding of science. Sociologists in the GDR and Eastern Europe had barely considered it. We became aware first of reports of surveys conducted in Great Britain and the USA,[1] and we thought that a comparison of the results of these surveys with results of similar surveys in the GDR could demonstrate interesting differences. In particular we wanted to know whether it is possible to demonstrate peculiarities that depend on the socio-political contexts of the shaping of public opinion. However our search for comparable polls carried out in the GDR was unsuccessful. We had been occupied with this subject only with regard to scientific administrators' understanding of science and its effect on research processes in the GDR.

This is why we decided to carry out a survey ourselves. We constructed a questionnaire and used it in East as well as West Berlin in order to gain answers to the question: to what extent did the socio-political system of socialism, which collapsed a few years ago, influence the public understanding of science? That is, the inquiry aimed to pinpoint the factors which have determined public understanding of science in the former GDR.

Before describing the results of our survey, we would like to explain why there might be grounds for assuming the existence of peculiarities we mentioned above. The traditional ideals of science which emerged in the nineteenth century as a resource for the legitimation of social politics was much more important in the former socialist countries than in the non-socialist world. The system was permanently praised as the materialization of a scientifically based socio-political programme. Scientific knowledge was attributed with the power to solve social problems. The tasks of establishing the new society were presented as parts of a central construction programme, the aims of which need not be negotiated in public – just like the aims of science. The presumed laws of social development were said to have a status like the status of natural laws. They were thought to be objective, necessary relations independent of the wishes and intentions of the people. Even these wishes, interests and so on were believed to be subject to an objectifying influence resulting from these laws of social development.

Because normative validity and the contingency provoked by public dispute about political programmes are mutually exclusive, it is not surprising that social development was thought to be a progressive process accompanied by a continuing reduction of contingency. In accordance with this the 'socialist democracy' aimed only to organize the participation of the people in the realization of a programme that could not in itself be a contentious issue. The people could not refer to regular procedures for critically reflecting the normative foundations of social institutions. We consider the separation of social politics from the shaping of public opinion to be one of the cultural side-effects of a social theory having developed into a 'science' under the influence of the objectivistic scheme of science.

More precisely: the objectivistic understanding of social processes is at first a result of the reception of the socialist concept. In *Die deutsche Ideologie* Marx and Engels wrote that the so-called objective historiography consists in the notion that the historical relationships were separated from the activity, and they characterized this notion to be reactionary. 'Scientific socialism' was understood in just this way by many Marxist thinkers and politicians.

The representatives of society's administration created the fiction of a rational society in which all determining social processes were scientifically described and regulated. Society's structure was also arranged in accordance with this structure. There was some correspondence between the way the GDR was organized and the technocratic reform of the socialist utopia. The complete scientistic-technocratic control and organization of the community pressed for a monist society in which knowledge and power are one and the same thing. The proponents of this society were inclined to favour a fordistic accumulation regime and held that the big industrial enterprises, understood as islands of rationality in the sea of anarchic market economy, promised the most hospitable milieu for a socialist economy and society. That is, we can relate the content of orthodox socialism influenced by the objectivistic understanding of science to the real organization of society.

We can imagine that the mentality of individuals who have lived in these conditions has been influenced in some respects. The extensive control of the community intervened in the course of individuals' lives in the name of socialism as science, i.e. a doctrine teaching the people how to live and what to strive for.[2] That is why we assume that conditions of life have influenced, at least in part, the GDR people's understanding of science.

Following the opinions of some authors who have reflected critically upon modern science, we believe that the public understanding of science should not be conflated with laypeople's familiarity with scientific insights nor dependent on their eagerness for broadening the mind. There are also important reasons for considering the socio-cultural conditions in which individuals formed their opinions about science.[3] We formulated the items for the questionnaire in dimensions expressing an objectivistic understanding of science versus an understanding of directions and orientations of science as a subject of public negotiation. We are not referring here to those older discussions of science which were limited to the analysis of the scientific language and the 'logic of research' but neglected the cognitive structure of science and the normative foundations of research. Logical positivism and critical rationalism assigned scientific rationality to Popper's third world, i.e. not related to cultural traditions. Both directions intended a radicalization of the codex of rational scientific activity and a strengthening of the shielding from influences of everyday knowledge. On the contrary, modern discourse, which refuses a dogmatically postulated reference system of scholarliness, is thought of as being open to the public's opinions. The resistance of scientific rationality to outside critique can be overcome if we reflect the cognitive structure in reference to a socio-cultural context.[4]

Now a few remarks on our survey. First we have to answer the question: can cultural groups, as opposed to individuals, be distinguished by their understanding of science? By contrasting the data about reactions of East Berliners with data about the reactions of West Berliners, we hoped to trace back the differences we found simply to the fact that Berliners formed their opinion of science within different political systems. Both groups share other socio-cultural factors: regional peculiarities, cultural background, and so on. Thus almost all third variables which might affect the understanding of science were matched save one central variable: the political system since 1945 and its various consequences.

We consider our inquiry to be only a first step for following the theme drafted in rough outline. We had to make methodological restrictions which mean that this is a pilot study: our survey was carried out by mail and the response rate was poor – we sent more than 800 letters to randomly chosen people, but received only 196 replies.

The participants were confronted with items which we thought expressed aspects of the understanding of science, and which were based on our study of the literature of science criticism. They included:

confidence in or scepticism of the utility of the science;
the significance granted to science in one's view of life or way of living;
acceptance of science as an autonomous enterprise or as an enterprise that is controlled according to socio-political aims and needs;
acceptance or non-acceptance of an epistemological privilege of science in relation to other cultures of knowledge.

We tested our assumption that people living in East Berlin understand science differently from people living in the West Berlin – in the sense above described – by means of discriminance analysis, a method for finding out whether the difference is significant between averages of many characteristics of two or more groups fixed a priori. In this way, we tested whether combinations of characteristics differentiated East and West Berliners with regard to their understanding of science. By means of this analysis, we also were able to eliminate variables which did not contribute to separating the two groups.

Using the data, we divided up the people questioned into two groups: group 1 contained cases with a more objectivistic understanding, and group 2 cases with a more critical one. Then we determined to what extent these groups are congruent with the actual groups (East and West Berliners). In the case of two groups there is the same a priori probability of $p=0.5$; i.e. by accident we would expect a correct classification of 50 per cent. But it turned out that this score was exceeded, as the table shows.

	Predicted group membership (to what extent are the groups fixed – statistically congruent with the actual groups?)		
	No. of cases	Group 1 (objectivistic)	Group 2 (critical)
East Berliners	88	76 86.4%	12 13.6%
West Berliners	47	2 4.37%	45 95.7%
Ungrouped cases	4	1 25.0%	3 75.0%

We found significant differences between East and West Berliners questioned with regard to the public understanding of science, which permits us to interpret differences with regard to objectivism versus acceptance of a public negotiation of directions or orientations of science. Our inquiry provided some reason to believe that East Berliners have more objectivistic understanding of science than West Berliners. They accept more than West Berliners the scientists' claim to be in a privileged epistemological position over other cultures of knowledge. They concede more readily that science is situated out of public discourse with regard to assessments and solutions of problems. Perhaps one can say that, in comparison with West Berliners' understanding, East Berliners' understanding is less connected with the present critical discussion of modern science.

We cannot say that West Berliners' opinion about science conflicts with the East Berliners' opinion. An objectivistic public understanding is widespread in all industrially developed regions. Here, we say only that we have statistically ascertained differences.

The results reported here possibly say something about one of the subjective reasons why broad sections of the East German population partly persist in subalternity and cannot adequately perceive new opportunities for creating a self-determined way of life or for influencing actively socio-political processes and institutions. Confronted with complications of the socio-political upheaval many people hoping for solutions to their problems possibly feel tempted to look for new authorities or 'objective truths' (such as the 'rationality of the free market') proclaimed by scientific experts. We believe that a reduction of excessive trust in science in favour of a more critical view is necessary for a revival of public discourse. The orientation to other authorities will not solve social problems. After disillusionment, chances will become apparent for the citizens to claim matters of the community as their own.

References

1   See Etzioni, A., and Nunn, C., 1976, The public appreciation of science in contemporary America. *Science and Its Public: The Changing Relationship*, edited by G. Holton and W.A. Blanpied (Cambridge, MA: Harvard University Press); Durant, J.R., Evans, G.A., and Thomas, G.P., 1989, The public understanding of science. *Nature*, **340**, 11–14.
2   In connection with this we find interesting the results of a survey on which authority is regarded as precious by broad sections of the Eastern German population. See Noelle-Neumann, E., 1990, Auf welche Unterschiede man sich einstellen muss. *Allensbacher Archiv*, IFD Umfragen 9002, Folie 39.
3   See, for instance, Meyer-Abich, K.M., 1988, *Wissenschaft für die Zukunft. Holistisches Denken in ökologischer und gesellschaftlicher Verantwortung* (München: Piper); Oettingen, G., and Seligman, M.E.P., 1990, Pessimism and behavioural signs of depression in East versus West Berlin. *European Journal of Social Psychology*, **20**, 207–220.
4   See Ginev, D., 1992, Varianten der kritischen Wissenschafts-theorie. *Journal for General Philosophy of Science*, **23**, 45–60.

Authors

Karlheinz Lüdtke studied sociology at the University of Leipzig, and works as a sociologist of science at the KAI e.V. Berlin, Prenzlauer Promenade 149–152, D-13189 Berlin, Germany. Renate Müller studied sociology at the University of Berlin, and is now based at the Potsdam-Kolleg für Kultur und Wirtschaft mBH, Wallstrasse 61–64, D-10179 Berlin, Germany.

Bridges to the future: scientific culture and conversion in Eastern Europe

James M. Bradburne

On 22 August 1991 all possible doubt about the extent of the changes to the Soviet political system had been swept away with the collapse of the attempted coup d'état and the triumphant ascendance of Boris Yeltsin as president of Russia. A new flag waved from the towers of the Kremlin, and a new era of instability and rapid and largely uncontrolled change had begun. By the beginning of the new year, the Soviet Union had effectively been splintered, and most of the former republics of the Soviet Union, including Georgia, Azerbaijan, Armenia and the Ukraine, declared their independence. The changes that had begun in earnest only in 1985 under the former president Mikhail Gorbachev had gained momentum and virtually swept away the political system that had been in place for over 70 years, a political system that had been considered a grave threat to world order since World War II, and that had been painted as the sworn enemy of the Western political philosophy of a democratically managed mixed economy.

The speed with which the former Soviet Union changed came as a surprise to nearly all observers, who believed that, lacking obvious successors, Gorbachev's stewardship would necessarily continue, and that the transition from the ideologically inflexible Soviet state would be gradual, and consequently coherently managed and uniform, taking advantage of the large and intact Soviet bureaucratic structure to build a bridge from one system to the next. The attempted coup of August 1991, the abduction of Gorbachev, the rise to power of Yeltsin, and the rapid dismantling of the Soviet state put an end to the notion that Russia's conversion to a Western-style democracy would be painless and gradual. The rapid unravelling of the Soviet system and the concomitant loosening of the political and ideological ties that bound together the nations of the former Warsaw Pact – the Velvet Revolution in Czechoslovakia and similar changes in Poland and Hungary – are central to understanding the term currently heard in policy circles in discussions of the new Commonwealth of Independent States: conversion.

This conversion has many aspects: the conversion of a state-controlled economy shaped since the 1950s by the necessity of maintaining military parity with the United States, the conversion of an inefficient industrial sector based largely on the production of military goods, and the conversion of an unwieldy infrastructure unable to meet the needs of growing markets, new enterprises and Western investment. This investment is often cited as the key to the process of conversion. In Russia, theorists of immediate change set in motion the privatization of large parts of economy, the floating of the rouble and rapid conversion to a market economy, for the moment largely unregulated. In less than a year, conversion has meant an inflation rate spiralling towards 3000 per cent.

Western imports to Eastern Europe are not limited to economic theories and investment, nor is conversion limited to the economic sphere. Until only recently, the informal science system of Eastern Europe and of Russia, comprising largely museums, was isolated by both economic and political barriers from new trends in museum design, planning and evaluation. Over the course of the past year, many of the obstacles to the free exchange of ideas have been swept away, and increasingly there is a perceived need to ensure that the museums of Eastern Europe are renovated, restored and revitalized. This can be called the conversion of scientific and technical culture, and is intimately bound to the educational and political structures put in place in Eastern Europe since World War II, structures shaped by the dominant Communist ideology of the Warsaw Pact countries. The structure of the informal science sector is also undergoing rapid change. In Russia, for example, many of the country's technical museums were under the aegis of the Ministry of Defence, and were thus hidden in locations such as Secret City No. 36, or Secret City No. 87, endowed with collections they were unable to display, in places they were unable to name. These museums of the Russian defence industries are now being allowed, for the first time, to speak of their history, of their collections, of their expertise. In the Czech and Slovak Republics, the museum community is being restructured, and old bureaucracies replaced with younger, more dynamic, and flexible administrations. In Hungary, the technical museums are beginning to make their voice heard in a community that for decades has put a premium on fine arts museums, and the use of fine art loans as a vehicle of cold war diplomacy.

The extent to which Western models of informal science education are being embraced can be seen by looking briefly at four projects currently under way in Eastern Europe. Each of them is indebted in one way or another to the science centre movement that was born in the United States in the 1960s, at a time when the formal education system was seen to be failing the next generation of Americans – a concern still very much at the fore thirty years later. In the West, the past three decades have seen museums begin to seriously re-examine their approach to the museum experience, and in particular the need to appeal to a broader public is acutely felt. Coincident with the need to involve a greater variety of visitors was the conviction that the approach found in older science museums was fundamentally flawed. Following the pioneering work in developmental psychology of Piaget and Bruner, and using their own research, Richard Gregory and Frank Oppenheimer began work in the 1960s on a new approach to the science museum, based on the conviction that learning science was best accomplished in a multi-sensory, 'hands-on' environment, where the visitor was encouraged to re-enact and rediscover scientific principles. Based on Frank Oppenheimer's principle of the three 'i's' – innovation, interaction and involvement – the Exploratorium became the model for a new generation of science museums, and has been widely copied throughout the world.

However, even in these new 'science centres' the visitors' experience was still largely defined by principles presented as discrete demonstrations, stand-alone and self-sufficient, despite the planners' initial intent to provide an 'integrative' approach. In the past ten years, a great deal of serious research has been undertaken to test the educational claims made by these new science and technology museums, and there is a growing consensus that there are yet better ways to excite the next generation about science and technology. New strategies in museum planning call for stressing the links between science, technology and culture, placing scientific knowledge in a social and cultural context, stressing the activity of the visitor in interpreting museum exhibits and creating new knowledge, and taking into account the competencies and expectations of the visitor in exhibit design.

What initiatives are the science and technology museums taking in Eastern Europe, and to what extent can they be linked to developments in the West? Since 1991, the National Technical Museum in Prague, Schola Ludus in Bratislava, the Museum of Science in Budapest, and the Moscow Polytechnical Museum have launched ambitious new programmes of renovation, including new exhibitions, new educational programmes, additional spaces and new buildings, which will each have to take into account the relationship between science, technology and culture, in the broadest sense.

In March 1992 the National Technical Museum in Prague opened its first 'hands-on' exhibition, 'Per experimentam ad scientia' (From experiment to knowledge), which was developed as a means of celebrating and explaining the work of Czech natural philosopher Jan Comenius. Many of the exhibits were inspired by those in Europe, found at the Palais de la Découverte and elsewhere, themselves often derivative of the models developed at the San Francisco Exploratorium. In April, as part of an ambitious renovation project, UNESCO and the EC funded an internal seminar with science centre specialists invited from France, Italy, Spain and the United States to share with the museum the Western science centre experience.

In September 1992, a group of university professors opened a temporary exhibition in Bratislava in the Slovak Republic, called Schola Ludus, coincidentally named after the classic by Comenius that inspired the exhibition at the National Technical Museum. The exhibition relied heavily on numerous staff explainers to demonstrate the exhibits, and was a huge success. It is currently being toured throughout the Slovak Republic, and is intended as a prototype for a fully fledged 'hands-on' science centre. Since 1991 the National Museum of Science in Budapest has been developing plans for a new facility, which they hope to inherit at the end of the World's Fair planned for Budapest in 1996. At present the collection of over 50,000 technological artifacts has no permanent home.

Russia, too, is about to embark on an ambitious programme of renewal, renovation and re-examination of its science and technical museums, which include several projects for new 'hands-on' science centres, children's museums, and the museums of the defence industries. As early as 1991, Novosibirsk had started to use interactive techniques found in Western science centres in the classroom, and established a small science centre. Researchers at the Institute of Culture are exploring ways in which to start a children's museum in Moscow, based on principles first elaborated in the 1920s by Zelenko, who made an extensive tour of North American facilities and published a long book on children's science museums in 1925.

In December 1992, to celebrate the one hundred and twentieth anniversary of its founding, Russia's largest and most prestigious technical museum, the Moscow Polytechnical Museum, opened Russia's first 'hands-on' science exhibition accompanied by explanations of scientific principles in children's verse. In order to benefit from work done in museums in Western Europe and around the world for the past twenty years, the Museum recently conducted an internal seminar so that Western specialists could share their experience with museum directors from throughout Russia. UNESCO granted technical support to the seminar and is contributing towards the publication of the seminar proceedings.

First of all it is important to underline that, given the speed and scope of the changes in Eastern European countries, there are as many differences as there are similarities between the different projects. In some cases the projects come from the educational research community, in others from the university, in others from the science museums.

Each country has had its own unique history, and even within a single country, or former country such as the Soviet Union, the changes that are occurring in all sectors are by no means homogeneous or uniform. In each of the four examples the initiative has borrowed heavily from Western experience and transformed this experience in light of their specific circumstances, giving a complex amalgam of borrowed ideas and innovation based on local circumstance.

Informal science is often seen in relation to the formal education system, for which it has traditionally played a supporting role. In Eastern Europe, however, before the 1917 Revolution, museums played an active role in education that was never fully matched in the West. In Russia, for large parts of the population, museums and excursions were seen as the mainstay of the educational system, with classroom instruction playing a secondary role. Nevertheless, despite this history of informal science education in which the Russian model was a world leader, Stalinist ideology put a heavy premium on traditional classroom education, reversing the relationship between the museum and the classroom. This renewed emphasis on formal classroom education became the hallmark of the educational system in the countries of the Warsaw Pact after World War II.

The emphasis on rigidly structured, streamed and ideologically coherent formal education was paralleled by a resurgence of neo-platonism in the sciences made necessary in part by the need to conform to the dominant ideology, although the exigencies of scientific research prevented its distortion to the same extent as the social sciences. This has left its mark in the scientific culture of all of the Eastern European countries and can still be seen in the informal science sector's dual bias, towards industry and industrial processes on the one hand, and with a 'soft', cybernetic and relativist approach on the other. Both of these tendencies can be discerned in the initiatives described above: the director of the National Technical Museum speaks of his museum's plans as symptomatic of 'a qualitative change of consciousness'; he speaks of the beginning of a 'World Age' and is committed to using the museum as a place where art and science can be fruitfully juxtaposed. The 'hands-on' exhibition at the Moscow Polytechnic 'Igrotekha' (Play/Technology) juxtaposes exhibits with rhymed children's verses describing the corresponding scientific principles, in a museum that celebrates Soviet industrial technology such as oil refineries and smelting facilities. Schola Ludus in Bratislava, in addition to interactive exhibits on resonance, insulation and conservation of momentum, also has exhibits on the greenhouse effect, the ozone layer and 'the relativity of our knowledge'.

Seen against this background, what conclusions can we draw about the strategies for building secure bridges from one form of scientific culture in the Eastern countries to another, and in particular about the growing acceptance of Western educational innovation in the informal science field?

First, just as Western economic aid is necessary in order to rebuild the battered infrastructure of the Eastern European countries as they meet the challenges of rapid economic change, Western expertise in areas of education and scientific culture is equally indispensable. Nevertheless, our experience in the Third World has shown us the limitations of the 'trickle down' effect, and underlines the fact that fuelling the economy alone is no guarantee that educational programmes will be strengthened. Economic forces are very slow to recognize the value of education, and in global terms capital is more likely to seek existing pools of highly educated labour elsewhere than commit long-term investment to its creation, or sustenance, in Eastern Europe. Eastern

Europe is lucky to have had an efficient and thorough formal education system, and as a consequence, a highly educated and well-trained population. Nevertheless, as economic collapse destroys the buying power of teachers and instructors, as social changes weaken the education system and as economic development is increasingly linked to short-term market considerations, the countries of Eastern Europe risk finding themselves with a severe lack of educated young people in the coming decades. It is to prevent a shortfall of scientifically trained, technologically literate and scientifically cultured people in the next decades that new initiatives in formal and informal science culture are directed.

Second, while Western expertise is necessary to stimulate new initiatives in scientific and technological education, borrowed Western models are bound to be inappropriate. Only now are Western science educators beginning to realize the active role played by the public, the importance of showing real science as it is practised, the central role played by context and the importance of culture to an understanding of technology. Many of these approaches already have deeper roots in Eastern European history than in the West, and there is a great risk that in embracing an educational philosophy coming from a specific but very different cultural, political and economic context, the countries of Eastern Europe will repeat mistakes already made in the West, wasting precious funds in the process. In making a jump out of the 1950s, the countries of Eastern Europe risk landing in the 1970s.

Third, remembering the past is indispensable to the creation and sustenance of a rich and varied scientific and technological culture, particularly when the past has been systematically distorted, withheld or erased as part of a broader ideological programme. The first science museums, such as the Arts et Métiers, played an important role in conserving a technological legacy for reasons that were not wholly disinterested: the challenge was industrial growth, and models and real objects were used to teach apprentice engineers the lessons of technology. The aim of science museums was to conserve in order to compete. They were active institutions in their social milieu: they were places to study, to teach and to display. They were part of a country's staying competitive and staying abreast of technological change. This remains a key role to be played by the informal science sector. In the examples described above, the initiatives already under way are strongest and most effective when they are rooted in the experience and history of their culture and their institutions, and weakest when they borrow unreflectively from already outdated Western models. The strong concern for the conservation of a rich technological history, a deep tradition of craftsmanship and an educational history that emphasized the importance of informal learning are strengths that should be reinforced and not discarded in the rush to catch up with Western innovation. But it is not enough just to enshrine the past. Museums must also play a role in creating a populace comfortable with, and capable of participating in, the technology of the future.

Finally, science policy is ultimately social policy, and long-term consequences can result from short-term decisions. Decisions in science policy shape the kind of economy that is built, the kind of products needed by the market, and the kind of professionals needed to service the market. As capital moves globally, decisions about the kind of scientific and technological culture that is fostered will determine whether investment flows into a region, and whether talented young professionals stay, or flow out. We have now entered a new era in Eastern Europe, a 'post-nationalist' era, and the expectations of the science education system are changing once again.

On the one hand, it is becoming increasingly clear that citizens want and need to understand technological change, in order to make informed decisions about events around them. In the light of concerns about the quality of the environment, the impact of continued industrial development, the Chernobyl catastrophe, and a history of being denied access to power and information, people are no longer willing to be merely receivers of information about technology, but are demanding to be active participants in evaluating this technology. On the other hand, the education system, from the classroom to the science centre, is under pressure to prepare young people to take up careers in technology. Increasingly society demands that informal science plays a role in encouraging young people to explore not only the function of technological artifacts, but also the creative thinking that has resulted in technological breakthroughs. Science museums must accept that it is no longer enough to teach young people about how an advanced aircraft works. It is not enough to explain the principles that make a banana-shaped fuselage. If Eastern Europe is to keep up with the West and avoid rapidly becoming merely a source of cheap labour, they must teach kids to think like its best creative scientists.

By learning from the mistakes we have made here in the West, and while daring to create new approaches to science education, married to their own unique history, experience and abilities, the countries of the former Warsaw Pact will be able to meet the challenges of the coming decades and create a vital and healthy scientific and technological culture rooted in their own reality.

Acknowledgement

This essay appears in part in *Planning Science Museums for the New Europe* published by UNESCO and the Národní Technické Muzeum, Prague.

Author

James Bradburne is an architect and museum planner specializing in new exhibition strategies for science and technology museums. He works with museums and science centres in Canada, the United States, Europe, the Middle East and Africa, and is based at 102 rue des Dames, 75017 Paris, France. He has been working in Russia and Eastern Europe for over six years, and is currently working with science museums in Russia, the Czech Republic and Hungary.

Science, Europe and the Third World

Mohamed Larbi Bouguerra

The aspirations, dreams, longings and reflections aroused by the progress of science in Europe have been expressed lyrically by the poet Salah Garmandi who wrote:[1]

> This Arab wants his proud and dignified ancestors, gifted with enthusiasm and passion, who in the past subjugated a large region of the world, and who assimilated and overtook great and remarkable civilizations by their creative genius. He wants those Arabs who were skilful tradesmen and great agriculturalists, who invented an irrigation system which gave vertigo to trees. He wants those men who could grow barley and corn which had the aura of miracle from their rather arid and barren soil, those men who were the first to know how to cultivate plants which would subsequently be grown around the world: the orange, the date, the apricot, the artichoke, cotton and sugar-cane.
>
> He wants his forebears, those who were great mathematicians, astronomers, chemists, renowned pharmacists and doctors, who could already operate without anaesthetic and had defeated measles and smallpox, talented architects who built palaces, mosques and towns, and created the Alhambra, Kairouan and Samarkand; those men who laid arabesque at the base of abstract art.
>
> The Arab of today, like the Arab of the recent past, has his memory; he remembers and reclaims the following words, which before being European were Arab terms: he wants cobblers (in French, *cordonniers*), tariffs, numbers (*chiffres*), cheques, he wants zero, zenith, nadir. He wants alchemy, stills (*alambic*), alcohol, amalgams, admirals, cables. He wants chain-pumps and irrigation canals. He wants algebra and algorithms.[2]
>
> But the Arab, reemerged from the past, wants the present. He not only wants his ancestors, but his contemporaries. He wants quanta, isotopes, transfusions, vitamins, turbines, electrons and elections. He wants gravitation and transmutation. He wants to master space. He wants the vaccine, insulin and penicillin. He wants chemotherapy, pre-stressed concrete, synthetic rubber. He wants nylon and television. He wants nuclear reactors, chromatography, the synchroton, cybernetics, and the antiproton. He wants sputniks, moonrockets, and satellites. He wants computers, jets, sub-machine guns, ground-to-ground, ground-to-air and air-to-ground missiles. He also, and above all, wants social justice.

In 1935, the barrister and Tunisian nationalist Tahar Sfar, addressing the leading European colonists in Tunisia half a century after the setting up of the French Protectorate, wrote:[3]

You talk about us as if we were a people only 50 years old, an infant people; but consider that on the contrary we are very probably an ancient people, and that we are suffering from being older than you, having already completed the entire life-cycle of a civilization. We had a culture which shone forth and which probably penetrated into yours, we had our men of genius, our creators. And though we are now fallen angels, we are angels nevertheless. We feel keenly the effects of our past which we carry within us. And you are badly mistaken, when, with your usual disdain, you take us for primitives, for the children of humanity.

Like an echo to Garmadi and Sfar, Abdus Salam, the Nobel laureate physicist, wrote:[4]

Science and technology are cyclical. They are the common heritage of all of humanity. East and West, North and South, all have participated equally in their development in the past, and will, we hope, continue to do so in the future, as our common efforts in science become a unifying force for the diverse peoples of this planet.

May God grant this wish! Joseph Ki-Zerbo, the Burkinabian historian of Africa, is rather doubtful. The exponential expansion over the space of only a few centuries of 'modern progress' – and in only a few countries – can produce neither change nor amazement. Two questions are enough to temper any scientistic triumphalism. First of all, who and what have been served by all these human exploits? Second, why do the risks for the very survival of the human race increase at the same rate as the breakthroughs in science and technology?[5]

Will Europe and her scientists pay any attention to these essential questions? Will they consign a certain 'eurocentralism' to a prop-room for historically out-of-date ideas? Will they recognize that their colleagues from the Third World ought to be helped as much as those from Eastern Europe who were fortunately spared colonialism? Alfred Kastler warned:[6]

Let us not forget that if we wish to put rational thought and scientific method to the service of mankind – be it a question of the physical sciences or the humanities – our aim can only be achieved if the fundamental motivation is a feeling of solidarity, in a word: love.

In the South, the concentration of means and the perspectives offered on the other side of the Mediterranean have the same effect as a light on moths. This 'brain drain' is the terrible scourge which the new Europe is likely to make worse.

The Sudanese scientific community – which is, alas, far from unique – is exemplary: the majority of professors, doctors, dentists, engineers and nursing auxiliary staff exercise their profession abroad. It is true that the lack of means (intentional on the part of the current administration in Khartoum), political discrimination and repression, and the absence of an atmosphere conducive to scientific work are also instrumental in this exodus – an exodus which some are only too delighted to see. Let the facts speak for themselves: between 1960 and 1990 Canada and the USA accepted more than a million managers and technicians from the Third World. Washington has saved 6 billion dollars in this way since the end of the Second World War.[7]

Will Europe have the imagination and courage to make bridgeheads of the scientists from the South which she employs, to make them links and mediators for new cooperation, for new scientific 'expeditions'; to study together, for mutual benefit, the richness of the South's biodiversity or the immensity of traditional knowledge? Will Europe have the necessary breadth of vision and will she agree to provide adequate funding for this adventure? Joseph Ki-Zerbo writes:[8]

> In African medicine there exist certain principles which have been valued, like the psychosomatic quality of certain products, and which have been neglected too much in Western countries. This shows that there are reserves of rationality, of logical principles in our African countries which should be exploited in the different branches of science to give new dimensions to these disciplines. It is here, I think, that we will find the best way forward, taking from Europe, and giving back in return. If we said that the Third World must start again from zero, we would be accepting an 'apartheid' of the mind.

'He who eats alone will choke alone', affirms the African adage. It is less a question of generosity than of interests which have been recognized. Today there are many, many questions which must be posed on a global scale, and it would be suicidal for humanity if the richest and most powerful retreated behind the walls of a fortress to which the destitute majority of the planet's population laid siege.[9]

In Albert Jacuard's well-chosen phrase, we live in a 'finite world', an interdependent world. A recent editorial in *Nature* reported that the town of New Orleans (population 500,000) is threatened by an epidemic of yellow fever, a tropical disease, and stresses that if an epidemic does break out, the stocks of available vaccine would be exhausted in a few days, and that it would then be necessary to import it from Brazil, where it is held in large quantities.[10] 'We always have need of someone less important than ourselves', as the virtuous Jean de la Fontaine said.

No-one can deny that progress in science and technology has served the poor countries, though infinitely less well than the rich ones.[11] It is no less true, however, that there are few scientific advances which have not been used with disastrous effect against poor populations to increase exploitation and dependence. The list would be long – from the first aerial bombardment in history, of Libya in 1913 by Italy, to the widespread use of herbicides (Agent Orange) by the English in 1952, against the Malaysian guerillas fighting for their country's independence. We need only evoke a few examples to clarify our ideas. Landsteiner's discovery of the different blood-groups opened the door to a flood of trafficking, revolving around blood and its by-products; a traffic of which Haitians, Indians and Iranians have borne the brunt. The discovery, by Professor Dausset, of human leucocyte antigen tissue groups had unexpected consequences: organ trafficking of all kinds. Did rich patients in London not have Turkish patients unwittingly operated on, so that they could then receive a transplant of one of their organs? In the Southern Hemisphere, is the unrestrained promotion of artificial baby milk, of antibiotics and of pesticides not an important factor in the mortality rate in poor countries?[12]

Could European scientists, following the example of the American group 'Concerned Scientists', not set up a watch-dog organization to denounce these intrigues and to help their colleagues in the South to detect products or equipment of dubious quality? Such an organization would also be inclined to help Southern countries to introduce technological changes, and to ensure that these countries do not miss out – or do not suffer on account of – the biotechnological revolution just as they missed out on the Industrial Revolution, because of colonization.

As for its practical applications, science appears to the South rather like Aesop's fables, surrounded by a halo of ambiguities. The production of vanilla by tissue culture will not make Madagascan or Réunion farmers rejoice; the enzymatic extraction of syrups from sweetcorn will appear like a condemnation to death for Filipino or Cuban farmers. Those of a dogmatic, intolerant or fanatical spirit find here a rich soil in which to sow the pernicious flowers of hatred and exclusion.

Who better than scientists from the Northern Hemisphere to help those of the South uproot these *fleurs du mal*, by spreading scientific culture and rationality on the one hand, and on the other, by preventing the profit motive from being the only thing to decide the fate of millions of men and women. From Marseilles to Algiers, from Manchester to Cairo, it is in all our interests to win this battle, for 'the womb that gave birth to the unclean beast is still fertile' – and that beast is heavy with Irrefutable Truths and Absolute Certainties.

In his day, Plato was already encouraging scientists to play a greater role in civic affairs. Human life has been transformed by science to a degree unprecedented in history, and Plato's invitation is still relevant in the case of the Third World. Furthermore, whatever the political solutions, the problems of the South will never be solved without the help of science. Did Francis Bacon not insist on the great practical value of scientific knowledge? The great ninth century Arab doctor El Assouli distinguished between the diseases of the rich and those of the poor. Today, he would have observed that the well-off part of humanity – a minority – is threatened by nuclear arms as well as by waste and pollution generated by the consumer society, while the poor majority of the human race is bent under the yoke of poverty and its resultant evils. But, as Abdus Salam notes, the Arab practitioner could find a common reason for all these negative aspects of the unequal world in which we live: on the one hand, science's excesses, and on the other the inadequacies of science and technology. Is it not within the capabilities of our European colleagues to reduce this gulf?

How? one might ask, and understandably so. By acting, for example, to reduce the monstrous registration fees for scientific conferences, insurmountable Berlin Walls which prevent us from participating in the 'High Masses' of our disciplines and worsen our isolation; by distributing up-to-date literature quickly to the South; by insisting that certain reviewers in the leading journals do not out-Herod Herod when they examine our manuscripts. And moreover, why not create fund-raising 'telethons' for countries at risk of disappearing from the scientific scene, to help them acquire laboratory equipment and products? René Maheu, former Director General of UNESCO, used to say: 'Development is science transformed into culture'. What an appropriate slogan for scientists from both the South and the North!

A heavy burden? Jules Romains replies: 'What is the point of living if it is not to correct our mistakes, to overcome our prejudices, and to open up our thoughts and our hearts?'

References

1   Salah Garmadi (1933–1982) was a poet and linguist; he held a teaching degree in Arts, and was Professor at the University of Arts and Humanities of Tunis. Among his many publications are

With or Without and *The Living Flesh* in Arabic, and *Le frigidaire* (The Refrigerator), which was published posthumously. He translated Saussure's works into Arabic, as well as *Je t'offrirai une gazelle* (I Will Give You a Gazelle) by Malek Haddad, *La Répudiation* (Renunciation) by Rachid Boujedra and *Moha le fou, Moha le sage* (Mad Moha, Wise Moha) by Tahar ben Jelloun.

2   For more vocabulary of Arabic origin, see Hunke, S., 1963, *Le Soleil d'Allah brille sur l'Occident. Notre heritage Arabe* (Allah's Sun Shines on the West: Our Arabic Heritage) (Paris: Albin Michel), or Vernet, J., 1985, *Ce que la Culture doit aux Arabes d'Espagne* (What our Culture owes to the Moors in Spain) (Paris: Sindbad).

3   Sfar, T., 1960, *Journal d'un exilé, Zarsis, 1935* (The Journal of an Exile, Zarsis, 1935) (Tunis: Editions Bouslama).

4   Salam, M.A., 1990, *Science, Technology and Science Education in the Development of the South* (Trieste: The Third World Academy of Science).

5   Beaud, M., Beaud, C., and Bouguerra, M.L., 1993, *L'Etat de l'environnement dans le monde* (The State of the World's Environment) (Paris: Editions La Découverte).

6   Droit, R.P., 1990, *Science et philosophie pour quoi faire?* (Science and Philosophy for What?) (Paris: Le Monde Editions).

7   *Le Monde*, 25 April 1992.

8   Zi-Kerbo, J., 1992, *Courrier de l'UNESCO*, February, pp.8–13.

9   Barry R. Bloom has defined the Third World as the part of the world in which 75 per cent of the world's population live; 86 per cent of births occur; and 96 per cent of infant mortality occurs. See Bloom, B.R., 1989, *Nature*, **342**, 115–120.

10 Editorial, 1992, *Nature*, **359**, 657.

11 In 1880, the ratio of per capita income in Europe and in Indochina was 2:1, by 1965 it had leapt to 40:1, and in 1991 it had reached a peak of 70:1. See the *Guardian*, 29 November 1991.

12 Bouguerra, M.L, 1993, *La recherche contre le Tiers Monde* (Research against the Third World) (Paris: Presses Universitaires de France).

Author

Mohamed Larbi Bouguerra teaches applied and analytical chemistry at the Faculté des Sciences de Tunis, 1060-Le Belvédère, Tunis, Tunisia. He has written extensively on scientific issues from a Third World perspective, and in particular on the impact of chemical science on developing countries.

The Basque country: science, culture and language

Alfonso Martinez Lizarduikoa

The Basque country, which sits astride the Western Pyrenees and is governed by two different states (France and Spain), is a small territory of some 20,000 square kilometres with a population of three million. The genetic characteristics of the Basque people seem to merge with its anthropomorphic and linguistic characteristics. On this basis, researchers accept as a working hypothesis the existence of a Cro-Magnon type human living in the area dominated by the Pyrenees, one of whose evolutionary descendants was the Basque who, some 4000 years ago, spoke a language with a structure very similar to that of today's Euskara, the Basque language. Thus the Basque language is the oldest surviving language of continental Europe, and a precious part of its cultural heritage.

Despite their genetic individuality, the Basque people have associated their separateness with their language and not with their race. In the Basque language, being Basque identifies with the learning of the language (euskaldun/esukaradun; Basque/a Basque-speaker). The fact of being Basque is therefore basically associated with culture. The Basque language was, furthermore, exclusively spoken until relatively late in its history. The first known written text in the Basque language is a document found in the monastery of San Millan de la Cogolla dating from the tenth century.

One of the periods of splendour of written Euskara was during the time in which trade developed in Lapurdi (Labourd, the French Basque province) in the sixteenth century. However, the economic, political and military changes experienced by the Basque people over the next four hundred years prevented its establishment as a unified political and historic entity (a territorial unity, with its own economy, administration, etc.), thus overturning any possibility of cultural unification centred on the language.

However, in the 1960s, during the economic crisis of certain sectors of the nationalist middle classes and the general crisis in the dictatorship, a strong nationalist movement emerged in the Spanish Basque country, the political and cultural influence of which still survives. The Basque culture then enjoyed a real resurgence and, in 1968, under the leadership of the eminent linguist Koldo Mitxelena, the foundations were laid for the unification of the language – a technical task which was entrusted to the Academy for the Basque Language (Euskaltzaindia). This period of linguistic standardization is not yet over.

Culture, science and private initiative

The Basque people have, over the course of their history, known a series of organizations which have operated at the same time as other institutions and which have, very often, given rise to projects which, in the final analysis and despite reservations, have been taken over by these other institutions. One example of this is the strong ikastolas (Basque school) movement which, thanks to private initiative, has resulted in over half the non-university student population of the Basque country being literate in Basque, using a totally Euskaldun model.

Another deeply entrenched movement was that of the fight against plans to introduce nuclear power stations in the Basque country. This battle resulted in a high level of awareness of the issue in vast sectors of Basque society, and finally completely paralysed the plans of the multinationals involved.

These alternative strategies have also developed in the context of science. However, since language is the fundamental vehicle for the communication of scientific ideas, these initiatives have come up against the recent standardization of the Basque language and against the failing of socio-linguistic standardization, which means that a large part of the Euskaldun population speaks but does not write the language. Furthermore, even among the literate, a large number of people who can write it shrink before documents of a scientific or technological nature, because of their reputation for unintelligibility – an unintelligibility which is further exacerbated by a language like scientific Basque, which is in the early stages of development.

Nonetheless, a group of professionals and university professors from various branches of science became aware of the urgent need to assimilate knowledge of the various sciences of the twentieth century into their own language, and formed a cultural society which they called 'Elhuyar', in memory of the Elhuyar brothers, who discovered wolfram, the tungsten ore (a discovery made in the Basque country in 1874 in the Vergara Seminary). The purpose of the society was to address the immense task of publishing textbooks and scientific literature, and a magazine which was to give coherence to the group. The magazine saw the light of day in 1986 as *Elhuyar Zientzia eta Teknika* (Elhuyar Science and Technology), and compares well with other scientific magazines on the market.

Publishing science in a minority nation

In this very individual context, there emerged in 1989 the GAIAK (Subjects) publishing project with an equally singular task: that of putting within reach of specifically Basque society the best works of twentieth century scientific thinking (natural and social sciences), with syntactic and semantic support in the form of a language which has come to us almost unchanged since the distant times of Cro-Magnon Europe.

The members of this project believe that a Europe with a historical and cultural heritage rooted in a plurality of cultures, ethnic origins and languages, and with its thousands of years of tradition, cannot turn its back on what constitutes its own blood without becoming estranged from itself. Every language or people on continental Europe which disappears from history contributes in some way to the death of this entity which we call Europe. Only from the reality of peoples can a cultural space be formed with a low water-line and with liberal perspectives. We think that the Europe of multinationals is dead as a cultural project. The Europe of major economic and military cartels is pushing us towards an alienated Europe which has little to contribute to the common heritage of the cultures on this planet at a time when twenty-first century Europeans are being forged.

In order to defend, cultivate and expand the culture of the peoples of the old Europe we must fight for the newly humanized European and the cult of the new century, for a people who are not one-dimensional, but who possess a range of registers derived from the history of the continent in which they live. The defence and the development of the cultures of the peoples of Europe also harbours a revolutionary potential in the sense that its very existence causes profound contradictions in the strategy of the multinationals which want an alienated people, with neither culture nor tradition, so that they can be controlled to serve their interests of the moment.

We think that people with deep cultural roots are better able to resist this manipulation on an intellectual level, as they would have experience and knowledge which they would want to protect and develop from a perspective closer to their own interests, and which, at any given moment, could place them in opposition to the interests of the imperialist machine. In our view, in a Europe where different cultures can be co-involved, can interchange and know each other, it will be more difficult to settle contradictions by means of wars of extermination, for which imperialism seems to have such a liking. We therefore think that by the sole fact of resisting cultural extinction, our people are helping to provide material of great value for the more open and progressive Europe of the future.

A small society such as the Basque society needs the science and culture of the twentieth century so as not to be by-passed by all the spheres of influence and action which in one way or another help to shape its future. In this way, therefore, the need to bring culture to our people via their own Basque language in all fields of science and culture seems to us to be a task of the highest priority. It was in this vast context that our GAIAK publishing project emerged. This project has, from the outset, backed the diffusion of knowledge as a means of socialization, by producing three collections of books. The first is intended for young people (from about 12 to 16 years of age) who are educated in the Basque language. This collection covers all branches of knowledge and all major subjects which interest our society, and is informative but makes no concessions; its content is modern yet critical, and sometimes even alternative. Among the subjects published to date are ecology, Islam, high-energy physics, sexuality, racism, the chances of extra-terrestrial life, the Big Bang theory, conflicts in Eastern Europe, the world of the subconscious and dreams, and much more.

The second collection publishes essays for adults and combines so-called universal topics (the conquest of space, the use of alternative medicines, the Darwinian theory of evolution and the theory of relativity, in the field of natural sciences, and also the role of the intellectuals in the Spanish Civil War in the field of social sciences) with others which analyse the specific technological and cultural responses our people have given to concrete problems (such as the development of technology and culture around whaling, Basque mythology and cosmology, racism in the Basque country towards the minorities such as gypsies, Agoths and Jews, daily life and socio-economic and cultural characteristics of medieval society in the Basque country, the independence movement and the origin of Basque nationalism, and also new ideologies relating to a progressive and internationalist nationalism which has emerged since the sixties, and so on).

The third collection is at university level. With it, we hope to make our contribution to a society which is equipping itself with the civilian tools it needs (although resources are still small) to take up the gauntlet thrown down by the new Europe without frontiers. More precisely, we hope that this collection will help us to respond to the challenge of university education in the Basque language, and of scientific and technological investigation and social science research in the Basque language in a fully international context.

This collection publishes in 1993 Jung's work *The Human Being in Search of his Soul*, Monod's *Chance and Necessity*, Weinberg's *The First Three Minutes*, Einstein's *My Ideas on the World*, Bonnassie's *The 50 Key Words of Medieval History*, Hallam's *The Revolution in the Earth Sciences* and more besides, which we intend to supplement over the next few years with works by Needham, Lakatos, Hawking, Thomm, Freud, Malinowski, Kandinski, Schrödinger and others.

The internationalization of science and culture

Our cultural project has as its starting point the fact that difference means richness. In this sense, we look to every culture, via its own specific features, to enrich the collective heritage with its particular vision of cultural phenomena. With regard to the Basque culture, our greatest current specific feature and our greatest potential for the future concerns our language. Today, the originality of Basque culture with regard to the language is reflected in its poetry (both sung and written) and its literature, which started to make its voice heard through the work of Atxaga. But how can science slot into this scheme of things? How can specific visions be produced in a system of knowledge which we can regard as almost global? How can any influence be brought to bear on a type of knowledge from which we are completely alienated and which is imposed on us by multinationals and the major centres of economic and political decision making?

A small people sustains its culture so that it can express itself or, better still, so that its reactions can be heard and taken into account as it designs both its own future and that of humanity collectively. This is a difficult task which each people that has survived the course of history has tackled in a different way. The Basque people, for its part, has maintained a dialectic relationship between isolation and openness. This attitude has fluctuated between self-protection, in which idiomatic, political and even mythological barriers were erected, and an attitude of openness to knowledge and the assimilation, into its own personality, of multiple contributions from an outside world which is, to a certain extent, hostile and aggressive. This attitude has probably been successful since, after thousands of years, here we are – still.

This same problem faces us today in the more general context of the Europe of the twenty-first century, under the shadow of a culture which standardizes and impoverishes current systems of North American mass communication. How can we meet this challenge when we do not control opinion-forming mechanisms and the 'culture' of the masses? We are positioning the GAIAK project at one of the poles of this isolation/openness dialectic mentioned earlier. We take up our position at the pole aiming to open the Basque culture to all the cultures on the planet, and so we must also accept that knowing and learning about others is a *sine qua non* for an ability to resynthesize and assimilate, in line with the historical and cultural references of our people, a universal learning to which we cannot remain immune without risking impoverishment and extinction.

With regard to scientific learning, to what extent can we speak of assimilation and resynthesis where scientific learning is today portrayed as a complete knowledge, in which there is no room for personal or collective interpretation or reassimilation?

On this point, in our view, science is an instrument which has political, ideological and cultural interests and ends, i.e. contingent historical references. In this sense, even if we do not have the wherewithal to create 'scientific knowledge', we can criticize and modify the scientific and cultural models to which we submit this knowledge. In this sense, we concur absolutely with the idea that the interaction between science and society is a two-way flow. This interaction is not always easy: evidence for this is the work carried out by vast numbers of our people who for ten years fought over the nuclear moratorium with the most fervent defenders of nuclearization – many of whom were nuclear engineers, respected representatives of political parties, economists and so on – who were beaten in the numerous public debates. Even when, as their arguments failed, they attempted to impose the project by force, it was civilian society which managed to paralyse one of the projects most dear to the multinationals involved, and whose prestige was at stake. Thanks to this process, Basque society managed not only to destroy the project, but also to become one of the ecologically most aware and best informed people in Western Europe. But this experience will have to be analysed in greater depth if the relevant consequences are to be drawn from it.

Moreover, a small people such as we are, standing on the threshold of the twenty-first century with a language which has been transmitted only orally and which must develop in all aspects to embrace the new concepts and ideas engendered by modern science, is in a perfect position to integrate into its language modern scientific concepts and to influence, via its own syntactic, lexicographical and semantic filters, a new world which will be new only if it manages to achieve this integration 'ecologically', in what constitutes its own system of signs forged over thousands of years.

To gain a better understanding of this 'reinvention' of science, I have chosen three very simple examples of the standardization which we are undertaking in the field of modern physics. Let us consider the concept of the space-time continuum, the semantic concept of which in English is roughly speaking 'the mathematical continuum which structures the variables of time and space in accordance with certain geometric rules'. This same concept in Basque has been defined by the expression *espazio-denborazko jarraikia*. Given that the *z* means *through the medium of*, *ko* means *of*, *jarrai* means *continuum*, the suffix *ki* means *made of* and *a* is the article, *espazio-denborazko jarraikia* is the semantic equivalent in English of, more or less, 'space-time via which the continuum is structured (made)'. There is therefore an undeniable semantic difference between Basque and English assumptions. Despite this, physics is no more difficult to understand in either Basque or English. In this way, we think that the genius and structure of each language colours the world with different chromatic scales, even though the realities considered are, to all appearances, as objective as the entities of science.

The second example is the electromagnetic field which, in English, is associated with the history of nineteenth century science. It could be interpreted roughly as 'the means of influence of electromagnetic forces', where we mean by *means* what Faraday and Descartes meant by it, i.e. full (rather than empty) space. In Basque we have elected to translate electromagnetic field by *eremu elektromagnetikoa*. *Eremu* is a much used word in Basque which, we should not forget, was essentially rural until the second half of the twentieth century. *Eremu* is a concept which in Basque defines an area of land which has limits or bounds. By semantic extension, we have given it the character of *jurisdiction* in everyday language and that of *field* in the sphere of physics. Thus *eremu elektromagnetikoa* expresses something like *jurisdiction where electromagnetic waves can act.* The difference between English and Basque is palpable. In English, 'field' means basically full (vibrant) space, while in Basque it is a 'legal jurisdiction (which can be formulated empirically) on which a certain type of forces act'.

Finally, I would like to cite the concept of the *quantum vacuum*. In English, the vacuum is associated with modern experimentation (nineteenth and twentieth centuries) in which the suction of a vacuum pump is applied inside a container, and this is accompanied by a fall in temperature which comes as close as possible to absolute zero. Then, below certain levels of space and time, fluctuations appear which can engender matter and anti-matter in very short spaces of time. What a huge theoretical content behind what is apparently so naive an expression! In Basque, the concept of vacuum is close to a paradigmatic word *huts*. *Huts* means *vacuum*, but at the same time it also means *nothing* or *not-being*. Absolute purity can thus have another semantic aspect. Now there are researchers who think that the Basque *huts* was the religious void engendered by neolithic people when they built their stone circles (cromlechs). This vacuum would be a wild and uncontrollable nothing which neolithic people controlled by art and magic. Our sculptors (Chillia and Oteiza) today continue this task of the emptying of matter, so that this vacuum can be examined on the basis of the problems and perspectives of the end of the twentieth century.

So for a Basque person the definition of the quantum vacuum as *huts kuantikoa* would be something like 'structure defined as nothing perceptible by human action (sculpture, architecture, music, etc.) and which is vibrant and not continuous at the deepest levels'. In this way, our people try, if this is possible, to keep hold of the thread which binds Cro-Magnon humans and their spirituality with the most advanced scientific thinking at the dawn of the twenty-first century. In our view, it is still possible to unify or, at the very least, to reconcile science and poetry, science and mythology and, finally, science and a history which is particular to each people.

It is in this context that we consider the GAIAK publishing project an architectonic creation, through which we can integrate the most modern concepts of thought into the neolithic structure of the Basque language, and with which we are attempting to plough an original furrow which will finally combine with the diverse contributions from other peoples.

There is no doubt: Europe's cultural heritage is precisely the diversity of its cultures. Knowledge of this diversity and reciprocal collaboration between them is the only way forward to a European unity founded on knowledge and plurality, and not on cultural impoverishment and monolithism towards which, it seems, today's hegemonic and economic forces are attempting to drive us. An initial step towards this European intercultural knowledge is the publication of this collection of essays which, we hope, will be but the first step in the exciting adventure which awaits us: that of offering future generations a more human science and culture which are intelligible to the individual.

Author

Alfonso Martinez Lizarduikoa has a doctorate in engineering, and is a lecturer in the history and philosophy of science at the University of the Basque Country. He also works for Editorial GAIAK, at San Bartolme 36 Behea, 20007 Donostia, San Sebastian, Spain.

Six stereotypes in search of a paradigm

John Ziman

I was seated in front of my word processor, toying with ideas for an article on the relationship of science and culture in Britain. Inspiration was dormant. A knock on the door woke me up. A slight, bearded, bespectacled man in a white laboratory coat entered, followed closely by a youngish woman, hand-in-hand with a boy and a girl in school clothes. They grouped themselves as for a photo. When the man spoke, I detected a provincial accent – Birmingham perhaps.

'I am Dr Boffin the Scientist,' he said, matter-of-factly. 'This is Mrs Boffin the Scientist's Wife. Here are Master Boffin the Scientist's Son, and Miss Boffin the Scientist's Daughter. We were meant to be a Happy Family.'

'Surely,' I replied, 'weren't we all! I am not a psychiatrist. Why have you come to see me?'

'We would like you to tell our story. The literary people don't understand what makes us tick, so we thought we should try someone who wrote about science ... Ah! Here comes the Prof.' Through the half open door came an older man, tall, clean-shaven, in a dark suit. The Boffins moved aside and looked towards him for guidance.

'Good morning! I don't think we have met before, although I know you well by reputation.' His public school manners were still fully operational. 'My name is Proff, Professor Proff. I am the leader of our little group, but I was slower than these young people in climbing out of the factory window.'

'Factory? What factory?'

'Let me explain,' he said, as he took the only other comfortable chair. 'There are six of us, and we have all escaped from the Innovations Centre of Party Games Ltd. They had put millions into developing, designing and demonstrating us. We were to be launched this Christmas. But then the marketing people discovered that we would not be competitive in Europe, so they decided to scrap us. We have only just managed to get away with our lives.'

'Well, that's fortunate,' I said. 'But what are you going to do now?'

'That's for you to work out', said Dr Boffin. 'We were only created to have a past. Our futures were to have been in the hands of the players. Of course my future would never have been very grand. My wife and I and our two children were to have replaced the family of Mr Coal the Miner in the old version of the card game Happy Families, but in the end they decided that Mr Tycoon the Currency Speculator was more modern.'

'Post-modern, dear,' said Mrs Boffin quietly. 'Those people in the marketing division with their masters' degrees in business administration are well into deconstructivism nowadays, even if they don't know a quark from a quasar. They think that discourse analysis optimizes their socio-economic roles.' 'Culture vultures!' sneered her husband, as I turned back enquiringly to the Prof.

'Oh yes, I was groomed to replace the Vicar in the latest version of that board game

Hoodunit?, the one where someone gets murdered in a country house. My wife can't stand country house parties, but I've always found them an agreeable way of making contacts. Mixing business with pleasure, you know.' He grinned almost boyishly. 'But you probably heard that they thought a Business Man would be a more suitable suspect nowadays.'

'You said there were six of you, but I count only five. Is Mrs Proff coming?'

'You must mean Lady Proff. They made me Sir Percival for my Report on Post-Graduate Training. I would have turned down the honour, but she loves it. No, she doesn't have enough secrets for a murder mystery. The person we are waiting for is called Schier Djeinious. I wonder what you will make of him!'

Like the devil of whom one speaks, in burst an arresting figure. He was short and stout, bald on top with a wild fringe of long white hair. The sparkling eyes and mobile mouth contradicted the aged, wrinkled face. His accent suggested origins in Central Europe: Ruritania, perhaps, before it was incorporated into Hungary after the First World War. But his vivid English was grammatically perfect.

'The bastards sent up a chopper after me. Those bloody people who wrote me into their new computer games are quite frightened of me now. They are very clever themselves, but they hadn't realized that a really brainy scientist who wasn't a villain would spoil all their scenarios. For a while I had to hide from their infra-red lasers. A pile of hay was much better protection, of course, than their black plastic armour, but some fungus spores have touched off my allergy.' He sneezed violently three times, then perched himself cheerfully on the corner of my desk.

By this time, I was hooked. 'So you want me to keep you all alive in a story,' I said. 'I shall have to know a lot more about each of you before I can start on that. Let's begin with the Boffins. They don't seem very complicated.'

I was right. Bill Boffin was a research scientist in an industrial laboratory. He had a chemistry degree from one of those civic universities in the Midlands, and had then taken a PhD in molecular chaotics at Camford. Now he was leading a small team working on hair conditioners. It was going very well. The new compound they had discovered five years ago had taken a large share of the market. I remarked that it didn't sound very interesting scientifically. 'Oh no, it's fascinating,' he protested. 'I've found I can use my old work in molecular chaotics. I think we're on the edge of a real breakthrough.' He tried to explain what he thought was happening, but I was soon lost in a maze of chemical structures and mathematical formulae. This was why, he said, he worked such long hours: after all, he had a much more rewarding job than the people in management, didn't he, even though they had all the power and got all the money. At home he watched football matches on TV and pottered around in their small suburban garden.

Mrs Boffin – Brenda – was actually a schoolteacher. She had been very bright as a child, and had gone to a girls' grammar school where she had not been discouraged from doing 'male' subjects like physics and chemistry. A scholarship took her to Camford University, where she duly got a First and was taken on to do research. But then she had married Bill, and what with two small children and a house to look after there just hadn't been the time to write up her thesis. Even if she had completed her PhD, she said, she would have found it very difficult to get back into research, after being out of science for nearly ten years. So she had taken a post-graduate certificate in education at the Polytechnic and was now teaching chemistry at the local Comprehensive School. Teaching was very hard work nowadays, with so much political pressure on the schools, but she was still enjoying it. The compulsory science in the National Curriculum was not very well thought out, but it was surprising how well it was being taken up.

Her two children, Lucy and Mark, were at another school, with a stronger academic tradition. They were doing very well. Lucy, now nearly 16, was about to do her GCSE examinations, and was expected to get top grades in all her eight subjects. Mark, at 13, was inclined to take things a bit easy at school, but he was already very interested in science and would probably work harder when he had to.

I turned now to Sir Percival. He, too, was not difficult to place. He was obviously a very big wheel in the scientific world. He had been Head of the Department of Molecular Chaotics at Camford University for ten years, and was a Fellow of the Royal Society. He served on the Science and Engineering Research Council and was Chairman of their Panel on Chemical Mathematics. He was also a consultant to several large firms and could often be heard on the morning radio demanding more government expenditure on science. He didn't actually say much about his own research, except to let it be known that he had a very large group of students and assistants working on some very important scientific problems, and that he spent a great deal of time flying to scientific conferences in Europe, the United States and Japan. I wondered about his earlier career. What had he done to deserve such success? It wasn't clear whether he was originally at Camford, or whether he had worked at any other British universities. We knew he was married; did he have any children? He was completely a public figure. I didn't see much point in questioning him about his personal life.

Finally, there was Schier Djeinious. Was he Dr Djeinious or Professor Djeinious? 'Oh, I would never have had the patience to write a doctoral thesis,' he giggled. 'But the University of Bessarabia gave me an honourary doctorate when I went back there the other day, and I'm supposed to be a Visiting Professor at Barchester, though I've never been to the place in my life.'

Whom did he work for? How did he make a living? 'I invent. I discover. I have ideas. Sometimes people insist on buying my ideas, but it's much less hassle just to give them away. Other people pay me for talking on TV or writing for newspapers. I enjoy that sort of thing so enormously that I would pay them for letting me do it.' He reminded us of his weekly TV slot, where he showed that seven-year-olds could understand the Second Law of Thermodynamics, and that seventy-year-olds could learn the Genetic Code. As a regular member of 'Eggheads' he had floored a famous theologian on the ethics of abortion, and a famous economist on the mechanics of monetarism.

'I'm not such a fake guru as I seem. My zany looks and weird accent are quite genuine. I overflow with erudition, but it's only good sense carried to the nth power. You don't have to believe everything I say. After I've said it, I often realize that I don't believe it myself. But I'm always worth listening to, especially when I talk about myself.' He giggled again at his own conceit. Did he have a wife and children? 'Oh, lots of both. Let me tell you how I did that research on baby talk. We believed in Piaget in those days, and wanted to see whether little Louise was keeping us awake on purpose. So ...'. I stopped him, for there could be no end to the private life that he was so obviously delighted to display in public. I addressed the whole group.

'Now that we know who you are separately, we should find out what you can do together. A story of action must start with an event. Suppose something goes really wrong in science. How are people like you expected to behave?' They were silent. Then

Mrs Boffin spoke up.

'Perhaps it's not relevant, but I've been very worried about Lucy lately. She has been having short lapses of memory, and her hair has been falling out. Could that have a scientific cause?'

Dr Boffin turned on her angrily. 'You're bringing up again your fantasy about it having something to do with our hair conditioner. I've told you a dozen times that it was widely tested before we put it on the market. That's not my side of the firm, you know, and I don't know anything about how the tests are done, but I'm sure they must have been quite thorough. They are very good scientists. You have to trust them.'

'Oh, they're excellent people.' The Prof broke in. 'I'm on the Board of the company now, and when there were some rumours of trouble with this product I had a good look at how it was tested. A few isolated cases of capillapathy were reported, but the symptoms varied from case to case. Statistically speaking, there is nothing to suggest that the hair conditioner was responsible. It must have been sheer coincidence.'

Brenda Boffin did not attempt to reply. She had obviously heard all this before. What soon threatened to become an embarrassing pause was broken dramatically by her daughter Lucy. 'Oh! Oh! Oh!' she cried. Her face became deathly white and she fell down in a faint. Her parents knelt beside her, and tried in vain to rouse her. The rest of us were frozen, stupefied, until Schier Djeinious jumped down from the desk. 'Let me help,' he said, quietly. 'I am, after all, medically qualified, if only from the University of Sarajevo.' After a short inspection of the patient, he advised that the girl was not in serious danger for the moment, but needed warmth and quiet. We carried her into the next room, and laid her on the sofa. I found a blanket and a hot water bottle. We returned to the study.

'This is extremely interesting', said Schier Djeinious. 'You say that there were a few isolated cases that were somewhat similar. What was the novel chemical ingredient, Dr Boffin?' Bill Boffin drew a complicated diagram with several interlinked hexagons and pentagons. 'Ah! Just as I expected. You can see that this is almost identical to the enzyme that promotes hair growth. No wonder it makes such an effective conditioner. But suppose somebody was born with a mutant version of the gene that produces this enzyme naturally. Then your compound might act as a suppressor rather than a promoter. The same enzyme acts on brain cells as well. That would explain the loss of memory. Of course, you specialists in molecular chaotics wouldn't have known about that, would you?' He could not resist this dig at his scientific colleagues.

'And how, may I ask, will you prove this conjecture?' said Sir Percival, icily.

'Oh, I shall need some genetic tests. It's just a matter of getting some blood samples from Lucy and her parents, to check up on her DNA at that site on chromosome 19. That should be easy.' Much to my surprise, Sir Percival blenched and blustered. 'Quite improper! You are not registered for research in that field. You will have to submit your proposal to the National Experimental Procedures Oversight Commission, of which, of course, I am Chairman. I shall make quite sure that it is turned down.'

'Steady on, Prof,' said Dr Boffin. 'I don't see the problem. There can't be any ethical objection to a harmless experiment to test such an interesting scientific hypothesis. And it might save Lucy's life,' he added, almost as an afterthought.

Sir Percival stood firm. There was an argument about whether the theory was plausible. Bill Boffin knew how difficult it was to make his former research supervisor change his mind. Eventually, he said, very quietly, 'I must insist, for my daughter's sake, that you agree to Dr Djeinious's plan. Otherwise, I shall have to write a very interesting little letter to *Nature*, or some other international scientific journal.'

By now, I was completely mystified. Why was Sir Percival standing in the way of such a test? What was it that Dr Boffin was threatening to disclose? As usual, Schier Djeinious had the answer. 'Ah, now I understand,' he said. 'It's really to do with Brenda. She was one of Proff's research students too, wasn't she? But because she was a woman he couldn't bear to think that she was a better scientist than he was. Also, she's a very attractive person. So he pushed what they now call "sexual harassment" to its extreme, and seduced her. She was going round with Bill Boffin at the time, so it was easy to cover up the fact that Lucy is really the Prof's child. No wonder he doesn't want a genetic test. The media would revel in this disclosure about the author of the famous report on Post-Graduate Training.'

Sir Percival's craggy face had turned to stone. But Dr Boffin cried out. 'Good God! I never realized it. Oh, Brenda!' For a moment they embraced. Then he produced another surprise. 'That wasn't what I was going to disclose – and he knew it. You remember that when Dr Djeinious came to England as a refugee he worked for a while with Percy Proff. That was just before Proff's first paper on chaotic molecules, the one that made his scientific reputation. Some people say that he ought to have got a Nobel prize for that discovery. Anyway, when I was at the Camford Laboratory doing my PhD I chanced on some old notebooks from that time. They showed that the original idea really came from Schier Djeinious. Even in post-modern science, plagiarism is considered much more wicked than sexual harassment.'

Schier Djeinious cackled with glee. 'Oh, don't worry. I balanced the account by having a lovely affair with the fragrant Emily Proff. Percival was so intent on cheating me scientifically that he didn't notice. It was a bit of a problem concealing the baby she was having while he was in America. It was a beautiful little girl, but I had no intention of hitching myself to the future Lady Proff, so we arranged to have it adopted secretly by a childless couple. I didn't learn anything about them, except that they were quite well off, and lived in some provincial town – Barchester I think it was.'

This time it was Brenda Boffin who cried out. 'Barchester! Did you say Barchester? That was where I was brought up. My Mum and Dad told me I was adopted, and that my biological parents had something to do with science. That's one of the reasons I so much wanted to be a scientist myself. You must be my real father. And you're my idea of a real scientist too.' She went over to him and hugged him. The old man beamed with pleasure. Sir Percival scowled. Bill Boffin didn't quite know whether to laugh or cry, so he began to explain to his son what they thought might be wrong with his sister.

Having no idea what to say next, I strode to the window, and gazed intently outside for a few moments. When I turned round, they were no longer there. I looked in the next room for Lucy, but she too had vanished. I was not altogether surprised. They had come to me in search of a story, and had found it, ready made, inside themselves. They had been created separately, for quite different roles, but had discovered how much they already had in common. What will happen to them next? I had woken from my dream, but they had become so vivid that I could easily imagine a plausible future for each of them. Schier Djeinious will luxuriate in being a grandfather, and will also become, through his daughter, one of those rare scientists who understand the real problems of science education. Sir Percival will not get a Nobel prize, but after chairing a Royal Commission on the Ethics of Gene Therapy he will achieve Lady Proff's ambition by being made a Life Peer. Dr Boffin will spend many more years as a backroom boy, before becoming research director of the company and discovering that he is not so bad at managing other people as he had supposed. Mrs Boffin would still like to go back into research, and is encouraged by Schier Djeinious to do a PhD in sociology. Through her book on the place of science and technology in national cultures she will gain an international reputation, and a chair of education at a prestigious university. Lucy Boffin will recover from her illness (which will never be satisfactorily explained) and will continue her brilliant academic career. But in spite of all the scientific talent she has inherited, she will wisely choose to study economics, and will become a very influential official in the Treasury. That leaves young Mark Boffin the Scientist's Son, still consumed by a passion for science. He too will achieve his ambition of a PhD in the new discipline of sub-atomic superchaos, but there will not be many jobs for people like that in Britain. After temporary appointments in America, Australia and Singapore, he will eventually be lucky enough to get a tenured post – in Paris, in the CNRS.

The deadline for my article is approaching, and I have written nothing. This account of my strange dream is all I can offer. But perhaps it does say something, indirectly, about the relationship between science and culture in Britain. How would such a dream turn out in Spain, or in Denmark? Look here! I'll try my hand at psychoanalysis and write an interpretation for the next issue of *Alliage*. It's 1993. The market is open: Europe is one – isn't it?

Author

John Ziman is Emeritus Professor of Physics at Bristol University, Bristol BS8 1TH, UK. He was Chairman of the Council for Science and Society, and Director of the Science Policy Support Group. He has written extensively on the social relations of science and technology.

1992: scientific culture in the New World

Goéry Delacôte

The three beautiful faces of science are: science as subversion of authority, science as an art form, and science as an international club.

Freeman J. Dyson

In order to better understand the various facets of scientific culture in Europe, it might be useful to attempt some comparisons with what is happening – or not happening – in this field in North America. I came to California from France less than two years ago, and so it is not too late for me to put forward some generalizations on the subject. I will do this now, then, before I become too well acquainted with it to dare to suggest generalizations. I will briefly analyse five determining elements in the emergence of a scientific culture: the role of the media, of scientists, of schools, of artists and of the science centres intended for the public.

Scientific information and the media

The first main difference from my French experience has to do with the scientific and technical news in the various media. In the daily press (in the *San Francisco Chronicle*, the *Los Angeles Times*, the *New York Times*, the *Wall Street Journal*, but also in the *San José Mercury News*) I regularly discover news on scientific and technical developments in the 'Business' pages. Of course, I also find it on the front pages (more often, it seems to me, than on their French counterparts), and on the inside pages along with general and national news. In the 'Business' section, the science is not simply the launching of a new product (the new incandescent lamp) or a new procedure (first the synthesis, then the almost parallel analysis of the therapeutic efficacy of thousands of molecules, thanks to the use of hybridizing, microelectronic and biotech microchips, which has made the recent success of the Affymax company); it is also failures, which are often linked to draconian, indeed paralysing, regulations (the Food and Drug Administration) or to a consumer market that is still hesitant about the technology in question (the pen computer). Scientific, technical, regulatory and financial topics, linked to companies or concerning particular personalities, intermingle freely on these pages. On the whole, it is a 'mixed culture' and not a separate culture, as I have too often seen in Europe. Although this may be a particularly Californian approach, it may not be insignificant in Massachusetts (Boston), Georgia (Atlanta), and now in Texas (Houston). Not all areas of science and technology lend themselves to this approach in the same way (a new space telescope, for example, might not), but this is not too important. For example, the defence of the 'B Factory', a particle machine hoped for by scientists at Stanford, is assured by their heavy reliance on economic arguments (local investment and employment, which is what matters in this time of elections) and technological arguments (existing know-how and findings), rather than by the knowledge that we could expect to gain from it on the matter–antimatter asymmetry at the origin of the universe.

On the other hand, specialist magazines which are well known to all scientists, such as *Science* and *Scientific American*, have a relatively limited circulation outside the scientific community. Like European magazines, they do not hesitate to tackle problems that have an ethical, social, political or economic element. Conversely, economic and financial journals, for example those that follow the trends in venture capital, are often explicit about the scientific and technical component of the corresponding innovation. In television, however, the spectacular, astonishing, illustrative and practical aspects will be highlighted, and much more so than on our channels (whether it is the cloning of dinosaur DNA from blood extracted from an insect trapped in amber, or the demonstration of the principle of an ultrasound refrigerator by the reverse experiment of attempting to produce sound in a metal tube from a temperature gradient).

The role of scientists

Let's start with an example. In September 1992, the Exploratorium, at the request of the school of medicine of the University of California, San Francisco campus (UCSF), was the setting for a twenty-four hour symposium intended for the general public and focusing on the scientific possibilities and the social inquiries arising from new DNA technologies. Organized by Harold Varmus, Nobel prize-winner in 1989 and a specialist in retroviruses and oncogenes, this symposium brought together some of the scientists who have made the history of the science of DNA: James Watson; Maxime Singer; Herbert Boyer and Stanley Cohen, inventors of DNA cloning; Gerald Fink, specialist in 'green' genetics; David Botstein, originator of the cartography of the human genome; Eric Lander, specialist in the human genome; Paul Berg, Nobel laureate ... . The history (even the anecdote about the role of a fast-food restaurant in Waikiki, Honolulu, where one night the technical idea of DNA cloning was born), the technology, the science, the ethics, the applications to health, to disease, to diagnosis as well as care (genetic risk), to agriculture and nutrition, the use of genetic 'fingerprints' as legal evidence – all of these aspects were reviewed, explained, and submitted to the scrutiny of the public and of the mediating journalists (from the *New York Times*, the *San José Mercury*, the *Wall Street Journal*, the *National Geographic*). The symposium was televized live to over ten sites (several campuses of the University of California, Stanford, etc. which were linked together by antennas and satellites) before a total of more than 5,000 viewers. Of course, we mustn't forget the speech sung by our friend Ira Herskowitz, expert in cell differentiation and president of the Department of Biochemistry and Biophysics at the UCSF. Herskowitz is almost as talented as the chemist Jean Jacques, who treated us, classical guitar in hand, to his 'Double Talking Helix Blues'.

The event was accompanied by continuous public access to the new experiments presented by the Exploratorium: the transformation of the blood cells of a patient suffering from sickle-cell anaemia (a genetic disease linked to one mutation), the movement in an electric field through an electrophoretic gel of bits of DNA made visible by fluorescence, the computer modelling of molecules (Silicon Graphics work station) and, a real puzzle of the transcription and the translation of DNA sequences into proteins, the study of ten genetic traits in each visitor and of the unique aspect of each individual, from statistics compiled by the computer. It is clear that this produced not only a science lesson but also a fierce debate on all sorts of economic arguments (the weight of the argument of world hunger in urging, or not, the genetic improvement of tomatoes), on access to information about the techniques, and on the ethical component and the stakes involved – and with significant intervention from the public. What emerged above all from the symposium is the limitation and the specificity of the scientists' intervention. A systematic questioning by the public of the scientists' approach showed that economic, financial, technical and social arguments are taken into account, but are generally handled less well than the technical aspects of the scientists' particular specialism.

This example illustrates the ability and the willingness of a large part of the scientific community to leave behind their own problems, methodologies and internal wrangling and to try to explain the state of knowledge and to react to a curious and often educated public. This exercise suggests an implicit culture and demonstrates the ties between scientific activity and democratic life – a very Popperian sentiment. In some ways large, interdisciplinary, scientific gatherings (the annual meeting of the American Association for the Advancement of Science, for example) afford some of the characteristics of such a symposium. Another example of the relations between scientists, the media and the public is the Scientists' Institute for Public Information (SIPI). The head office of the SIPI is in New York, though it now has offices all over the country, and its council includes many prestigious scientists. The task of the SIPI is to assist the media by providing lists of scientists willing to respond quickly to journalists' questions. France has recently organized an analogous mechanism, following the English, under the auspices of the Académie des Sciences. In particular, the Institute tries to acquaint the United States with the role of foreign scientists (Japanese, European, Asian) and, naturally, comes up against the difficulties one would expect, given the often dominant, even dominating position of American science.

Scientific culture and the schools

Science and mathematics education, from kindergarten to the final year (Grade 12), is in a state of advanced decay in American schools. The reasons for this are known: there is a shortage of qualified teachers; science curricula are sometimes 'optional' and intermittent; teaching is bookish and unappealing; and there is a lack of sensitivity and considerable inaction on the part of the larger framework regarding the suspicion with which individuals from certain family backgrounds view intellectual procedures. I should also mention the contrast between the diversity of ethnic cultures in the US (in California the current majority of European origin will be a minority before the end of the century) and the membership of an overwhelming majority of science teachers to the so-called 'Caucasian majority'. Strangely, the main difficulty is not, first and foremost, a financial one, since the spending per student has risen by 40 per cent in real dollars in the past ten years (what a striking parallel with France, at least in this area!). One characteristic of the American educational system is that it is completely the responsibility of individual districts (there are 16,000 academic districts in the USA). In particular, the kind of lessons to be taught and the recruitment of teachers are for the most part determined at the district level, or even at the school level. The result **is** a very disparate science education: 30 per cent of high schools do not teach physics at all. In terms of science education, therefore, it is a *local culture* that develops, rather than the infinitely more normative, more integrative and, all things considered, more national culture that develops in European countries.

Such is the magnitude of the disaster that it has caused real shock in the USA. From teachers to politicians of all sympathies, through the scientists and the industrialists, a new flexibility – a flexibility which has never existed before, since it is contrary to the country's traditions – has given rise to a national agreement in favour of the promulgation of national standards with regard to the content of courses, the methods of evaluation (something a little less rudimentary than multiple choice questions), and the appropriate methods of teaching. Certainly, it is not a question of programmes, which would be to swing sharply to the other extreme of excess rigidity, but rather of guidelines for choosing what will be taught and for negotiating *all* of the related requirements (training, learning environment, and so on). The effort began in mathematics three or four years ago, and I am fortunate to be involved in the reform of science education on the national level. At the same time, attempts at management innovations are blossoming with the creation of new schools and a new approach focusing on the student, the use of technologies (creation, communication, information processing, access to knowledge bases with processing more or less incorporated), and a new use of time, of teaming formats (groups of various sizes), and of the role of the teacher (facilitator, negotiator of goals and meanings, cognitive guide, etc.). It is too early to judge all of these initiatives. The more general context is pierced through with tensions and contradictions, particularly with the rising multicultural approach which does not necessarily converge toward an integration of minorities, but rather toward a highlighting of their specificities and their differences. Any overly optimistic or pessimistic predictions would be out of place. I have the feeling that new practices will emerge, but for the time being, Grade 12 science teaching is not doing at all well.

The artists get involved

From time immemorial, some of those who are traditionally called artists have maintained relations with the scientific activity of their period. Sometimes they borrow the scientists' vision of the world, their manner of representing time, space and knowledge, their tools or their technologies. In the other direction, there is lurking at the heart of every scientist a taste for form, for elegance and for the aesthetic. Although this taste does not always have the opportunity to express itself, it is in large part at the origin of the powerful moments and emotions of scientists who recount their paths to discovery. Science and art share the privilege of being able to look at the same world but from new angles. On the subject of scientific culture, I also see extensive differences between France and California in the role played by artists. One could perhaps describe the Californian approach as 'de-complexed' culture. And today's research, with its new fields of investigation (materials, turbulence, chaos) lends itself more and more readily to this approach.

There are, of course, artistic performances that borrow from contemporary technology. A show by Georges Coates, in a church transformed into a theatre, presents actors acting as much in front of as behind (at several levels) a semi-transparent screen.

This screen is capable of showing still images, animated or mixed projections, or 3D images of the real world or of a world reconstructed on a Silicon Graphics work station (the latter makes possible most of the special effects of the latest, most talked-about movies such as *Terminator* and *Born to Die*, thanks to the talents of the engineers at ILM, one of George Lucas's companies). Another example is the revelation of the beauty of molecular forms, and the representations of chemical reaction processes, in an exhibition of collages by the chemist, Roald Hoffman, Nobel laureate, and his colleague, Vivian Torrence. A third example is that of the invasion of space by a troupe of acrobatic dancers, the Zaccho group, against a background of music composed and improvized with the aid of numerous computers – a sort of anti-gravity aesthetic. But it is also a new way of perceiving the astonishingly beautiful architectural structures of the Hall of the San Francisco Palace of Fine Arts which houses the Exploratorium, and which was designed in 1915 by the non-conformist Californian architect, Bernard Maybeck. All of these experiments are about incursions and borrowings. They express an audacity. The results may vary greatly, but this is in the nature of experimentation. In my opinion, what constitutes the originality of this artists' approach to science resides in their ability to select a natural or technical phenomenon and to present it in a way that respects its complexity, facilitates its exploration and underlines its aesthetic qualities. The meanderings of water trickling down an inclined surface of dry glass, the double funnel of a water-vapour tornado, the dynamics of sand dunes shifting with the wind, the dance of rings of cold vapour formed by the depression of a disc of soft material – these are some examples of an aesthetic created by artists from scientific phenomena. In this respect, an approach founded on the illustration of concepts would be an impoverishment. The chosen approach allows, particularly at the Exploratorium, a genuine intervention by the artists, and not simply the decoration of the exhibition.

Toward new institutions

Up to this point, I have not mentioned the role of science centres – in France, one would say 'scientific, technical and industrial culture centres'. First, an observation: they are mushrooming all over the USA but also throughout the world, even though it is a difficult time everywhere because of the recession which diminishes both 'earned' revenue (admissions) and 'raised' revenue (contracts, donations, support from municipal, state or federal authorities). But let's look a little more closely at what might constitute the difference between a US science centre (the Exploratorium, for example) and its classic European equivalent (the most recent centres, like the Experimentarium in Copenhagen, tend to follow the example of the Exploratorium). In brief, I would say that the US approach is more favourable to the emergence of an inquiry based on exploratory actions, these being inspired by the choice and the design of the element of exposition. Inversely, the European approach favours more an access to knowledge *illustrated* by the element of exposition, and it is understood that, in the end, the visitor will finally have accepted the concept put forward by the creator.

In a sense, we find here the two sides of our understanding of the world, even if, in practice, there is never a perfect fit between the two conceptions. This explains why the second approach, the European, favours historical and social contextualization to bridge the phenomenon–concept gap and to make it accessible, while the first will be more strongly determined by the nature of the 'slice' of world presented by the subject of the exposition, in the here and now, to the curiosity of the public. It is thus a more exploratory culture that is expressed by the 'Explo' approach. There is always the risk of seeing nothing at all, of missing the point, but especially the risk of not exploring the canonical routes of knowledge and therefore of not progressing in the knowledge of reference. On the other hand, however, the exploratory culture has the advantage of developing its own capacity for inquiry. Of course, this penchant for risk-taking will be maintained by the assurance, for the visitors, of being astonished by unexpected behaviours or even by cognitive clashes when they interact with the phenomenon. There is an implicit reference to common knowledge, an important component which intervenes in the art of the presentation of the phenomenon. There is, in addition, a yet unexploited possibility, from the technological angle, of cognitive contrast that could lead to a double interaction: that of the individual with the phenomenon, and that of the individual with other individuals considering the same phenomenon – a reflexive approach.

The future of these institutions of scientific culture probably belongs to those who will bring together the two approaches, and allow the visitors to venture a direct, unpredictable contact with the phenomenon *and* to navigate through existing knowledge by following their own line of questioning. These future institutions will offer, to a wide public, *opportunities* to create discovery laboratories where the tools for the exploration of the phenomena and for multimedia navigation within the fields of knowledge will be elaborated by the same people who want to use these tools to understand and to learn. It is possible that such institutions, bringing together several functions – those of libraries, museums, practice centres, open laboratories, media centres – could play an ever more important complementary role in relation to and in the service of existing institutions (a school, a specific medium) by helping them to open up and to evolve. The key is in the diversification of functions, which could be elementary in the fight against ossification and immobility.

These cultural and educational schemes, distributed but also linked together, may be an important innovation for the beginning of the twenty-first century, and will thus contribute to the strategic evolution of colossal sectors such as academic organization or extensive media (such as educational television stations). In effect, they offer many advantages. First of all, they are a crossroads for people of different abilities and disciplines; they are also reception areas for individuals, families and groups, proposing very diversified kinds of activities (independent, self-guided tours, work sessions, more extended training sessions, etc.) with various agendas, from recreation to training. As the centres or the heads of networks, they permit a diversification that combines a contact with each participant based at his or her home, and a socializing and meeting place to bring them out of their isolation, thus ensuring an inter-individual dimension of the creativity, the communication and the learning. By calling upon technologies with which their users will experiment, and adopt or reject, to bring into play certain of their functions, these organizations will permit delocalization and will lead to new methods for managing time, space and resources – skills that will be profoundly different from those we know today.

It must be hoped that our French tendency to fragment cultures and to separate practices (science, management, economy, finance, administration and artistic creation are seldom mixed) will not be too great a handicap to us in the emergence of these new institutions. The non-sophisticated, pragmatic, experiment-oriented side of American culture could favour them. On the other hand, the American obsession with the short term, the absence of collective memory and the juxtaposed ethnic cultures are potential obstacles. There also, we are in mutation. It will be exciting to follow the evolution of this movement and to see if it will contribute to the entrenchment and the enrichment of a better integrated scientific culture.

Conclusion

In this brief panorama of the actors in a scientific culture, I have described the appearance of cultural styles differentiating the United States from France and, more generally, from Europe. Mixed and implicated cultures, local culture, de-complexed culture and exploratory culture are expressions of this difference. Obviously, differences within the United States can also be considerable; this discussion is not meant to be a detailed nor an exhaustive analysis. In particular, nothing is said about the 'anti-science' cultures, about the various forms of eugenics, sociobiology, creationism, etc. These movements have, or had, a relatively limited influence. The first official framework for the teaching of sciences in California (the California Science Framework), recently made public, contains an unambiguous reference to evolution. This decision does not appear to have led to the same effects in the ultra-conservative clan as did the video of the treatment inflicted on Rodney King by police officers in the poor neighbourhoods of Los Angeles.

All things considered, this analysis, though brief, does reveal that the two continents have much to learn and to gain from their respective approaches. And for this there is absolutely no need, thank goodness, for a new Christopher Columbus.

Author

Goéry Delacôte is Executive Director of the Exploratorium, 3601 Lyon Street, San Francisco, California 94123, USA. He is also Professor of Physics, currently on leave, from the University of Paris. From 1982 to 1991, Dr Delacôte was Director of Science and Technology Information, one of eight scientific divisions of the Centre National de la Recherche Scientifique (CNRS). He is a Board member of the International Academy for Education.

A European's private understanding of science and technology: a personal mini-history of dilemmas

Marcel P.R. van den Broecke

I was born in the Netherlands while World War II was raging, but fortunately remained personally unaffected. There were still food shortages when I was young, but life was visibly getting better each year as I grew up. My first conscious acquaintance with science and technology was made at an exhibition called 'The Atom' when I was twelve. It showed that atoms could not only be employed to destroy Japan, but could also be applied peacefully to generate energy that would be too cheap to meter. The catalogue of the exhibition quoted Albert Einstein, the undisputed authority on scientific matters, as saying 'When we succeed in applying the discoveries of nuclear physics for peaceful purposes, the road to a new paradise will be opened'. Since I was eager to enter paradise, I applied all my youthful curiosity to increasing my knowledge of this microcosmos, and became fully determined to contribute to it by becoming a nuclear physicist. On the macrocosmic scale, similar vistas were being opened. The Soviets launched their first satellite, Sputnik, weighing an impressive 53 kilograms, and I collected all the newspaper articles I could find on the subject. They predicted that Mars would be reached within ten years, that control of space would end any necessity to control the Earth, that the Moon would be mined, and I was excited and grateful to be living on the brink of an age when the human race would be finally in control of nature, to the benefit of all, including me.

On the more intimate household scale, I witnessed the first refrigerator being carried into the house, an ingenious machine with the capacity to generate cold out of heat. A vacuum cleaner ousted the laborious carpet-beater, a washing machine the wash-board, a spin-dryer the wringer, a dryer the clothes-line, an oil heater the coal stove. My mother and I marvelled at these advances in technical ingenuity which no longer strained her back and liberated time that could be devoted to grander issues than dirty laundry. This was the time of my life.

My favourite book was Semjonow's *The Riches of the Earth*. It listed and discussed all the known and hidden natural resources of the Earth, and particularly of Siberia – resources that lay waiting to be explored, just there for the taking, to be put to use by humankind. Imagine what a wonderful place the world would be when people applied their ingenuity to exploring and utilizing all this potential! I did imagine, and liked what I saw. Science and technology meant truth and progress. Who would not like that?

In this high spirit, I constructed radios, first with a crystal, then with a vacuum tube, and finally with transistors. I set up a chemical mini-laboratory to make colourful fireworks, crystals and stench bombs. I constructed elevators, gearboxes and differentials with Meccano, and thus achieved my own private conquest of nature.

Western Europe was a quiet corner of the world at this time, recuperating from its wounds, struggling to get back on its feet. But America and the Soviet Union – those were the countries where the action was. Since America made its presence felt economically, militarily and, most of all, culturally in Western Europe, America would be the obvious road to paradise. And anyway, Holland was too insignificant to be chauvinist about. So America was where I went at the age of eighteen to study science, as soon as the opportunity arose. I arrived in the summer and received a crash course in English and, most interestingly, in American civilization. It had never occurred to me that such a subject could be studied, and this conviction was strengthened by the contents of the course, which dealt with such unlikely subjects as the influence of the emergence of the American car on teenage necking and petting behaviour. In my view, if this was to be defined as civilization, it was not an object of study, but something that formed part of yourself in a long chain of traditions that developed over time. The teacher of the course reacted to this view with a strange mixture of distaste and sympathy – distaste for my lack of understanding that America was a young country, with an attitude of 'make do' that left no room for my subtle misgivings, and sympathy for my display of scepticism so typical of the Old World. I should cultivate my foreign accent, was his advice. Too bad it was not a French accent: that would have increased my attractiveness to the opposite sex. But a Dutch accent would also do, indicative as it was of my European background, steeped in culture and tradition which indeed, as he conceded gracefully, was lacking here. I returned to Europe, trained in English, science, and American civilization, and feeling more civilized than I had ever felt before.

My curiosity and fascination with science and technology persisted as I continued my studies in Amsterdam, majoring in linguistics and human speech. Computers had entered the research scene, and could be applied to these subjects as well, with fascinating results such as mechanical translation which was expected to be operational soon. I absorbed as much of my scientific surroundings as I could. In short, I built up my scientific literacy.

Wealth increased rapidly in Europe. The big German cities Bremen and Hamburg were no longer the shambles I had seen before on my cycle tours. I bought my first car, and so did many others. The huge material gap between America and Europe began to disappear. But so did my visions of paradise and the belief in my capacity to contribute to its realization. I felt invaded by the uneasy feeling that my increase in scientific and technical knowledge did not bring paradise closer to me, or to anyone else for that matter, but rather seemed to push it beyond the horizon. I wondered what lay at the root of this feeling, what it was that caused my dream to crack.

Nuclear energy was not too cheap to meter, and turned out to produce nuclear waste, hazardous to me and many generations to come, and reactors presented their own risks. Yet, as each house I lived in was bigger than the previous one, my energy consumption grew, in spite of the application of better insulation and economy lights. Each car I had was more reliable and longer lasting than the previous one, but also bigger; and I was spending more time idling in ever longer traffic jams. My cars generated their own contribution to acid rain and the greenhouse effect, and finally to the growing garbage dump.

Space research turned out to hinge mainly on an arms race, and was even too costly in that limited respect. There was nothing to mine on the Moon, no priority for Mars, no extraterrestrial life to guide us. We seem to be stuck on this planet, and to be making a mess of it, and the universe does not seem to care.

The population of the Earth has doubled in my lifetime, and if my life expectancy holds and trends continue, I will live to see another doubling. That has never happened before, and it will probably never happen again, but it is no cause for joy. Advances in medical science have made this possible, but they turn out to be a mixed blessing. I am happy to owe to medical science the fact that both my parents are still alive today, but I wonder how my children and grandchildren will fare in a world they have to share with too many fellow passengers. These passengers will no longer stick to their own poor, crowded part of our global village but will come to any place where food is still to be had. I see them starving every day on the TV news which so conveniently comes to my living room while I am digesting my well-balanced meal.

I have come to know what habits and what foods cause cardiac and vascular diseases, bone weakening and cancer, and have given up trying out all the medical advice, some of it contradictory, that is poured over me through the media. It gets on my nerves.

I enjoy the cool drinks and fresh food from my refrigerator, but I know that the CFCs from the seven refrigerators I have used up since my youth have each contributed to the ozone hole that is widening rapidly so that the Sun's ultraviolet rays may hit me or my children sooner or later and burn us to death. Yet, I do not stop using refrigerators. My behaviour does not change, but my feelings about my behaviour do.

What is to be done with these examples of the dubious rewards for fostering my scientific literacy? Am I alone in my frustrations? Should I, as an emotional victim of lost innocence, crushed between knowledge and intentions, long for the blissful ignorance that characterized previous generations of happy, scientific illiterates, and try to reclaim it?

A well-known Dutch psychologist, popular for his easy solutions to complex problems, attributes my contradictory behaviour to the reptilian part of my brain, which usually wins if at loggerheads with my rational neocortex. This is occasionally unpleasant, but is nothing to worry about – it's only human. This explanation, if it deserves that name, does not appeal to me, because it denies the fundamental nature of my dilemmas.

And not just my dilemmas: the pros and cons of science and technology sit side by side, and the internal conflicts they generate are everywhere around me. Scientifically literate people in Europe, notably the north-western part of it, seem to be more aware of these conflicts than their counterparts in America or the Far East, the other two parts of the world where science and technology have taken hold of society.

Most Americans seem to take it for granted that their dominance of nature, and their greedy consumption of natural resources, not just from their own country but also from many others, is the natural contemporary version of a divine right, reserved for God's own country. Even where awareness of pollution has led to severe legislation, such as in California, cars remain big and petrol cheap.

The Far East demonstrates its awareness of these issues which accompany the increase in prosperity in that part of the world by showing sound levels and pollution concentrations at street corners in big digital displays. It pays for the enjoyment of material wealth by living in cramped quarters, commuting for long hours in overcrowded trains, working itself to death – in short, by enjoying a low quality of life.

It seems to me that Europe has the potential to lead the way in coming to grips with the vicissitudes of science, technology and industry which, after all, originated here. We cannot achieve that by turning back the clock; there is no way to unscientize or detechnologize society. If there is no paradise around the corner of the future, neither is there a past to which we can return.

Just as science and technology have provided the tools for making the products and circumstances that have led to the present dilemmas, I believe that scientific and technical tools must be developed to help us find our way out. The first step in developing such tools is the insight that science has no ultimate truths to offer, but only a variety of options. Decisions concerning which road to pursue of all conceivable roads are made by human beings, not by mysterious outside forces or by technology as a blind, driving force by itself.

To know what choices we face requires scientific literacy, but that is a means, not an end in itself. Why might a proliferation of scientific literacy be expected to help in solving problems induced by science and technology? Because choices of how to use them then become a collective responsibility, rather than a personal one: on a personal level, I am willing to take the train and relinquish my car if I know what the benefits of this attitude are, and if my neighbour does the same. I am interested in buying a CFC-free refrigerator if other refrigerators are taxed out of existence, or no longer produced. I am willing to finance more food for the Third World, if effective birth control measures are implemented to make it a more meaningful gesture.

If a scientifically more literate community, to start in Europe, is the prerequisite for working towards a goal of collective responsibility for an integration with nature, rather than its subjugation, I am willing to exert myself in making my contribution to bring this about. Of course, given the immensity of the task, this means little. A collective effort is required.

Fortunately, many organizations and their staff and members contribute to an infrastructure to help implement this intention. On a European level there are, for example, PCST, the international network on the public communication of science and technology, and ECSITE, the association of science centres; these interact with national organizations such as the Foundation for Public Information on Science, Technology and the Humanities (PWT) in the Netherlands, and its counterparts in other countries. Lacking the strength to solve my dilemmas alone, I am happy with doing my share of work for PWT, in the hope that through teamwork I may contribute to solving my personal dilemmas and regain some perspective on my lost paradise.

PWT was founded in 1986 as an initiative of the Dutch Departments of Education and Science and of Economic Affairs to stimulate and improve public understanding of science and technology in the Netherlands. It receives an annual budget of 2 million ECUs (5 million guilders). The aim of PWT is to function as a centre of expertise for promoting public understanding of science, technology and the humanities. Its activities include advice, new initiatives, coordination, financing and conservation. Thus, it intends to function as a node, a source of information, a partner and a facilitator in the national Dutch network of actors involved in public information on science and technology.

PWT offers advice to intermediaries for setting up, implementing and evaluating activities to stimulate public understanding of science and technology. Knowledge and expertise based on experience with its own projects and on research in communication studies can be helpful in this respect. It undertakes new initiatives (such as Professor Pen Pal and Science-by-Phone), and addresses new target groups (such as youngsters aged 8–12, inside and outside school). New themes are introduced (such as biotechnology) and new media explored (science theatre, travelling exhibits, science festivals).

PWT stimulates coordination between actors developing initiatives in the field of public understanding of science who share target groups, aims or methods. This is achieved by promoting contacts and exchange of materials and experiences. The monthly PWT newsletter *Iota* provides information to those active in the field on these topics. PWT also gives financial help to initiatives in the field of public understanding of science (such as TV programmes on science and technology, museum exhibits, folders and publications, meetings, youth programmes and so on), and provides funds to maintain programmes that have proved their quality and effectiveness in promoting public understanding of science and technology such as the National Science and Technology Week, or youth clubs for young scientists or astronomical observatories for the lay public. Thus, PWT develops initiatives of its own and supports those of others. Through this dual function, PWT is in a position to develop expertise on public information about science from two different angles, as an initiating and as a facilitating/evaluating agency. This benefits both activities.

Any information effort which only stresses the advantages or disadvantages of developments in science and technology is doomed to fail. Nowadays only a realistic presentation of the pros and cons stands any chance of being credible and acceptable to the public. PWT has no specific interests to promote, except the spreading of information. Therefore, it tries to inform the public in a balanced, pluralist manner without trying to influence it towards refusal or acceptance. If the public still opts against certain scientific and technical developments, then such a refusal, if based on arguments, is much to be preferred above refusal, or even acceptance, on emotional grounds: arguments tend to provide a more solid basis for discussion than uninformed emotions.

Acknowledgements

I am grateful for constructive comments by Lou Dalderup, Esther de Groot, Marten Knip (PWT) and Deborah Günzburger.

Author

Marcel van den Broecke is Director of the Foundation for Public Information on Science, Technology and the Humanities (PWT), Nicolaas Beetsstraat 218, 3512 HG Utrecht, The Netherlands.

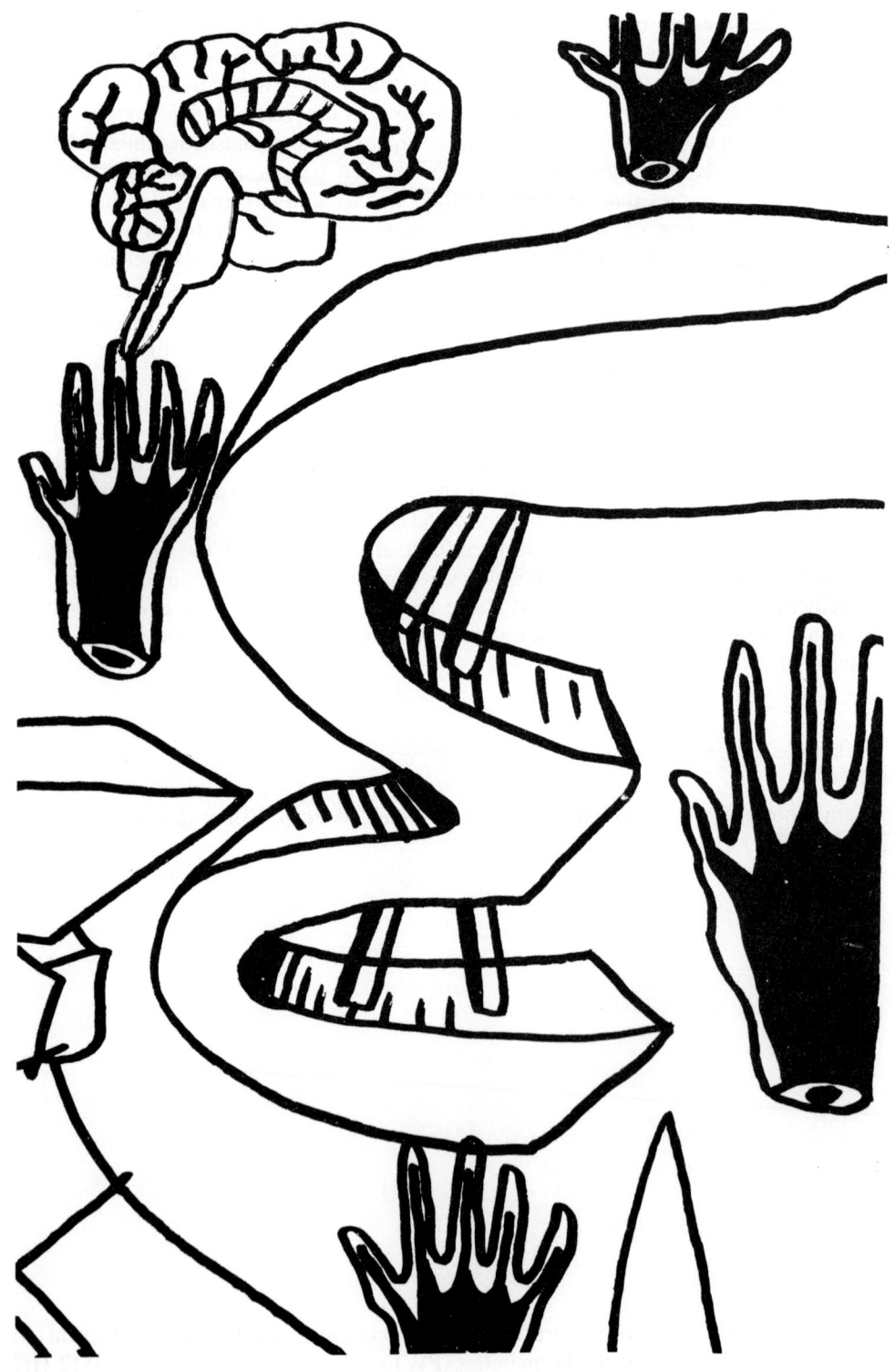

Hypotheses

Science and thought in Europe

Dominique Lecourt

Modern science in Europe was born of a collective effort which, allied to powerful economic expansion and the invention of new political forms, ignored national borders; if in England, as early as the beginning of the seventeenth century, Francis Bacon sounded the final charge against the last Scholastic, Galileo, and was familiar with his works, Galileo took the first decisive step in Italy when he started to mathematize the motion while confirming the heliocentric hypothesis formulated by Copernicus in Poland. In France, Descartes opened up new horizons in terms of this mathematical reasoning by creating 'analytical geometry'. Leibniz in Germany then took over and founded the basis for infinitesimal calculus while proceeding with an intensive critique of Cartesian metaphysics. Without the work carried out at the same time in Holland by Christian Huygens, a disciple of Descartes, on the movement of pendulums, Isaac Newton in England would not have been able to achieve the process of constitution in 1687 without encountering increased opposition from the Continent in the form of the Leibnizians.[1] From this point modern science proved to be international. It drew from the rich body of work produced by Arabic and Indian scientists, in particular with regard to mathematics and optics, and resembled the Greek tradition, which itself was associated with the scientific traditions of Egypt and Babylon, as it was rediscovered in the twelfth century AD. Today, like yesterday, science knows no borders.

The focus of basic scientific research shifted to the United States of America shortly after World War II, notably thanks to the contribution of numerous European 'brains' – Jews and democrat exiles from Nazi Germany and those who came subsequently, attracted by the better working conditions. The Manhattan Project, a research programme formed with the aim of manufacturing the first atomic bomb, provided an organizational model which has since become established in the scientific community worldwide. The procedures and specifications of scientific research have become standardized. Managed and administered according to efficiency objectives which are uniformly quantifiable in terms of experimental results,[2] science has essentially been conducted ever since in another world of thought.

This world wanted to turn its back on England at that precise moment when modern science was developing rapidly. It retained an empirical vision of science from the 'old Europe' which it combined with a renewed 'natural theology'. Result: the philosophical approach adopted by European scholars in the vast debate on science, and which later formed the intellectual stimulus for its most outstanding progress, never really found any adherence in the United States. Alexis de Tocqueville was right when he observed that 'nowhere else in the civilized world is there a country where less attention is paid to philosophy than the United States'.[3] History has for the most part only confirmed this diagnosis. This is gradually being realized on the other side of the Atlantic during a period some people describe as 'post-analytic'.[4]

It is true that America did devote itself to an original philosophy, that of 'pragmatism' based on the work of William James (1842–1910) after it was passionately acknowledged in the maxim of Ralph W. Emerson (1803–1882): 'rely on oneself'. However pragmatism and its initial inspiration did not resist for very long the success of 'logical positivism' which was introduced by Austrian and German émigrés including Carnap, Reichenbach, Feigl and Hempel. This was followed by an epistemological professionalization of philosophy which, cut off in highly specialized university departments, devoted itself solely to the technical solution of formal problems of logic and language without establishing a link to living scientific research. This withdrawal proved propitious to the expansion of a silent philosophy, a nameless philosophy which organizes the cult of 'fact', 'progress' and 'efficiency' around the mythicized image of Francis Bacon. The social sciences and humanities, the development of which maintains a direct link to social demand, have given substance to this silence – a silence which, it is true, has been pierced on occasion by a few methodological discussions, and yet which has never questioned the generous presuppositions accepted by these disciplines. Such a silence represents a constant opposition, which may be latent or open, between a certain triumphal technology, and a reaction of a deep-seated popular hostility towards science in the form of religious values. The misadventures of Darwinism in the United States cannot be explained any other way; 'scientific creationism' promoted by Protestant fundamentalism appeared to be the retort to the 'materialism' flaunted by the sociobiologists at the end of the 1970s.[5]

If one looks at the conditions which fostered the last two outstanding revolutions in the physical and life sciences, one discovers a Europe at its most intellectually effervescent, the Europe of the second half of the nineteenth century which examined the philosophical basis of the successes and failures of Newtonian physics and which, using several linguistic registers, applied itself to formulating a vocabulary which was transferable beyond national borders. The works of the greatest philosophers, in particular of Kant, together with Berkeley, Hume and Plato, formed the subject of an intensive, meticulous and passionate analysis. From Vienna to Berlin, and from Berlin to Prague, London and Paris, it can be seen how Ernst Mach, Ludwig Boltzmann, Heinrich Hertz, James Clerk Maxwell, Henri Poincaré and many others adjusted those categories from which the theory of relativity and the quantum theory were drawn: 'real', 'reality', 'objectivity', the representation *(Darstellung or Vorstellung)*, model *(Model)*, image *(Bild)*.[6]

The same can be said of the life sciences. Molecular biology was by no means indebted for its birth in a single stroke to the quantum revolution in physics, and brought into being by Erwin Schrödinger when he published *What is Life?* in 1944.[7] In reality it had been under preliminary study for a long time in the form of debates in biology prepared by Ernst Haeckel (1834–1919) in the steps of Darwin, by August Weismann (1834–1914), then by the Belgian Hugo De Vries (1848–1935), one of the 'rediscoverers' of Gregor Mendel (1822–1884), and, in a more decisive way than one might have believed, by Claude Bernard and Louis Pasteur in Paris, contemporaries of the author of the *Origin of Species* (1859). The philosophical propositions by Leibniz, Kant, Schelling and Comte on the conceptual relationships of life have been put to the test and support the belief of one and all when confronted by the question of questions: that of the finality of nature.[8]

With regard to the social sciences and humanities, the theoretical bases of their progress remained in Europe until recent years when they became the subject of open questions. As strong as the attraction of empiricism has been, and as respected as it could be, and rightly so, the so-called 'fieldwork' and debates in France, England and Germany which, according to divergent directions, marked the birth of sociology, have been ceaselessly revived by different phenomenological schools (the Frankfurt School, structuralism, and so on).[9]

Admittedly there is not a 'European science' of the type which Edmund Husserl (1859–1938) heroically believed he had to restore to its ideal in 1935 in order to save a certain concept of humanity which was facing danger – but there is a European scientific concept and practice which had always planted a passionate philosophical argument in his heart and which governed the relationship between scientific knowledge and other conditions of thought and action.

Today this European style finds itself in danger of being lost under the empire of planetary technology. A descriptive account of the knowledge and accountable practice of research may be found in most laboratories and teaching establishments. The knowledge and practice of research may be magnified by audio-visual means which favour only the most spectacular results, and which are used to spread enthusiasm or to sow terror. Its self-justification as an instrumental concept of rationality inspires all the devastation of the social environment of human life, to the detriment of those populations whose living conditions are the most primitive. The risk associated with this research is great if our world reacts to this unforeseen disaster by indicting researchers, deifying nature and making sacred what is 'natural' in people. We know that this ideology has caused suffering and the death of millions of people in the recent past.

Are we going to see a return to the trench warfare in which the scientific devotees of science and the obscurantist anti-science crusaders indulge?[10] We fear that the current spectacular expansion of the parasciences – the technological forms of popular superstition such as computerized astrology, crystal ball quantum therapies, numerology, etc. – will trouble people.[11] This much praised 'new age' could, when stripped of its manufactured mysteries, become a nightmare; the scientific certainty which has been placed at the service of tyrannical sects would not be slow to fanaticize those people in distress who hand over their destiny to their 'masters' – people who are unscrupulous when it comes to the profits. The Church itself has good reason to worry.

Has it not become urgently necessary to re-examine the philosophical purpose of scientific thought, and to release it from its positivist straitjacket? The teaching of sciences should first be reviewed in this context. Instead of aiming to inculcate results and utilitarian conclusions which are always uncertain, why not try to transmit the basic essentials: the sense of the problem, the determination of the unknown as a possible object of knowledge? It is already almost half a century since Gaston Bachelard expressed his wish for this type of 'training of the scientific mind' which would reintegrate the learning of equations and theories into the conflicting history of audacious thought to which they owe their birth. This reorganization would offer the additional advantage of reopening the decisive question of the future of human existence and the correct pace of technical thought, which does not confine itself to the 'application' of scientific knowledge any more than it constitutes the unacknowledged essence of science. If this idea was encouraged by social, political and 'biological' values, it would liberate us from current technology and the technophobic humanism which opposes it.[12] The media could be invited to cease taking the easy way out, without falling into the trap of didacticism.

Here, at least, is an inheritance whose value could reunite European intellectuals and restore the splendour to a mutual school of thought with a universal orientation, which would be based on open argument subjected to properly controlled experimentation, and offer an opportunity to revive technical ingenuity and stimulate the creation and rationalization of diverse forms of life. This new impetus would protect thought from the fluctuations which for the last half century have led alternately to the celebration, not at times without presumption, of scientific triumphs declared guaranteed in advance, and to the announcement of the apocalypse for tomorrow which science itself has brought. There is no doubt that the words of 'old Europe' invoke a great echo in the United States among all those who, with courage and lucidity, have tried to identify the intellectual motives of the cultural crisis of which essentially they, together with their fellow citizens, are both the witnesses and the victims.

Acknowledgement

This essay was published in *Le Monde Diplomatique* in February 1993, and taken up again in *A quoi sert donc la philosophie?* (What is Philosophy for?) (Paris: Presses Universitaires de France, 1993).

References

1   For a scholarly analysis, see also the classic works of Alexandre Koyré and the recent book by Blay, M., 1992, *La naissance de la mécanique analytique* (The Birth of Analytical Mechanics) (Paris: PUF).
2   In particular, see the analyses of Rouban, L., 1988, *L'État et la Science* (The State and Science) (Paris: CNRS).
3   de Tocqueville, A., 1961, *De la démocratie en Amérique* (Democracy in America) (Paris: Gallimard), Vol. II, p.13.
4   See the articles collated by John Rajchmann and Cornel West entitled *Post-Analytic Philosophy* (New York: 1985) (and in French as *La pensée américaine contemporaine* (Paris: PUF, 1991)).
5   Lecourt, D., 1992, *L'Amérique entre la Bible et Darwin* (America: between the Bible and Darwin) (Paris: PUF).
6   See the well-documented presentation of Niels Bohr's book *Physique atomique et connaissance humaine* by Catherine Chevalley (Paris: Folio, 1991).
7   The French translation by Christian Bourgois was published in 1986.
8   The influence of German 'Naturphilosophie' on biological thought has been described by Gusdorf, G., 1985, *Le savoir romantique de la nature* (Romantic Knowledge of Nature) (Paris: Payot) and analysed by Klein, M., 1980, *Regards d'un biologiste* (Through a Biologist's Eyes) (Paris: Hermann).
9   Lepenies, W., 1991, *Les Trois Cultures* (The Three Cultures) (Paris: Editions de la Maison des Sciences de l'Homme).
10 The recent Heidelberg appeal seems to be the bearer of such a threat: see the articles in *Le Monde Diplomatique* by Ignacio Ramonet (June 1992) and Jean Marc Lévy-Leblond (August 1992).
11 See Terré-Fornacciari, D., 1991, *Les sirènes de l'irrationnel* (The Sirens of Irrationality) (Paris: Albin Michel).
12 Simondon, G., 1969, *Du mode d'existence des objets techniques* (The Existence of Technological Objects) (Paris: Aubier Editions Montaigne).

Author

Dominique Lecourt is President of the Association Diderot and Professor of the Philosophy of Science at the Denis Diderot University (Paris 7), 20bis bvrd de la Bastille, 75012 Paris, France.

Science, Europe and democracy

Tullio Regge

The tormented relationship between science and society has become fashionable, and is taking the lion's share in countless meetings and symposia all over the world. The debate about science and research is raging in Europe, and members of the European Parliament are in a privileged position to observe what happens. Unfortunately, they are rarely listened to by the people who take the real and important decisions.

Modern science was born in Europe, but for reasons which are as yet unclear we have lost our leadership, and we now rank third behind the United States and Japan, and perhaps even behind Taiwan and Korea in applied research. We are still leaders in the field of elementary particles because of CERN, a prestigious institution; but CERN cannot be considered as purely European, and it is not in any way connected to the EC.

In assessing the present situation and forecasting future trends we must take into account the special and dual role of science in modern society. Science and research are the main factor behind the spectacular technological evolution and economic growth of industrialized countries of the last two centuries. Science is also an intellectual endeavour and part of modern culture. As such it is a social force which is far more powerful than we dare to admit.

There are no sharply defined boundaries between these roles, and this brings me to some of the points which I would like to discuss more at length. I remember quite clearly that when Hiroshima was destroyed by a nuclear bomb some historian predicted that the present age would be called the age of the atom. Nuclear energy became the symbol of indefinite technological development, and once more we were swept by a wave of triumphant positivism. Oppenheimer declared that physicists would never be unemployed, and the general feeling was that we were heading toward the ultimate technological and scientific society. Almost fifty years later the public image of science is tarnished and it is quite clear that basic irrationalities in the behaviour of the masses are still quite powerful and rampant.

I am convinced that science is under attack, and from various directions and with different purposes. I do not have enough space to analyse this impressive phenomenon which covers a vast and incredibly varied spectrum, and which includes at one end highly regarded philosophers such as Croce and Heidegger, and on the other end of the spectrum the nature buffs, animalists, shamans and the like. Suffice it to say that in pursuing any future research policy we must come to grips with this reality, and not simply dismiss it as background noise or a mere annoyance. The roots of popular opposition to official science go deep into the past and may produce unwanted results.

As a member of the CERT (the Commission for Energy, Research and Technology of the European Parliament) I listen as a matter of routine to speeches and reports of colleagues and of representatives of the Executive Commission. Nobody seems to care very much about the public image of science, and in fact the kind of decisions and advice which we are usually requested to handle has to do with technical details of research programmes and rarely with principles. The Framework Programme is by far the most ambitious scientific and technological programme ever launched by the Community, and in spite of its shortcomings it had and still has a deep and beneficial influence within the EC. I should clearly state, however, that this programme views science solely as the primary technological propulsive engine; there little room for culture or for pure science, and even less for the public image of science. This stand is in part a consequence of the famous – or rather infamous – Subsidiarity Principle and of the need to revive Europe's technology, which is now sadly lagging behind the United States and South Asia.

On the other hand, the popular image of science will ultimately become an important factor in determining the outcome of elections, and eventually of our research and technological policy. Quite recently a deputy in one of the leading Italian parties was elected on a pledge to stop all public vaccination programmes. The need to restore a sane and rational understanding of science in the public should be obvious, but it is not yet dealt with in any appropriate way.

But naturally we are also interested in knowing what to do with our limited resources as we enter in the third millennium. It is now clear that fundamental research in the field of elementary particles is getting increasingly costly and that the most high energy physicists can hope for is a single super-accelerator to be used worldwide. Purely theoretical physics has no future, and illustrious examples of the past show that empirical evidence is badly needed to keep it going. Perhaps the most spectacular example is the near total failure of Einstein's efforts towards a unified theory. Either physicists discover some new and revolutionary ways to make cheaper and better accelerators, or high energy physics, as we know it today, will wilt and disappear. Alternatively physicists will discover that the biggest natural accelerator which ever existed is the cosmos itself, and will try to derive high energy data from cosmic rays and cosmology. This brings me to the forecast that astrophysics still has a long way to go, and that it will absorb the talents and resources that would have gone to high energy physics.

On several occasions the CERT has hosted discussion on the future of nuclear fusion. Many colleagues have hotly debated the advantages and liabilities of this form of energy in a future world economy. I think, and many experts agree, that fusion will not be a significant factor before the year 2050, and my impression is that many of the debates on the safety and convenience of fusion border on the academic. In any case, we are dealing with a field in which the technological problems clearly overshadow the intellectual challenge.

I come finally to biology, and to the problematic field of bioethics. I am frankly disturbed by the excessive emphasis given to bioethics. There should be no separate morality for biology, or for that matter for science, and I do not see the need to open a separate chapter for it. I see a continuous and often spectacular evolution of biology and I admit that some of the new techniques may create unforeseen problems. We must nevertheless refuse and fight against the stereotype of the biologist as a creator of monsters and as a reckless individual who spreads deadly diseases. We must accept the fact that the path of science can be a dangerous one, but at the same time it is worth following; indeed, we cannot turn back. We are facing tremendous worldwide problems which cannot be solved simply by stopping and burying our heads in the sand.

By the years 2010–2020 most of the current drugs will be totally ineffective against strains of drug-resistant bacteria, and new diseases may appear following the blazing trail of AIDS. On a planet where the population goes on growing unchecked and even encouraged, a planet swept by famine, by desertification and vast climate changes, there is little room for optimism. My feeling is that we are condemned to research whether we like or not. The alternative is chaos and a ghastly return to the medieval plagues.

If science were to disappear its role would be taken by a heterogeneous army of magicians, shamans, crackpots and religious nuts far worse and more dangerous than the Dr Strangeloves they are supposed to replace. In order to avoid this we must change the public image of science, look anew at the problem of popularizing it, and explain honestly and clearly the purpose, limits and dangers of research. R&D activities in the Community and elsewhere are hit by a crisis of vast proportions, and the Community must pay attention to the immediate problems of revitalizing them. In the long term we must also take care of science as a form of culture and as the ultimate human drive toward deeper knowledge.

Author

Tullio Regge is a Member of the European Parliament, where he serves on the Committee for Energy, Research and Technology, and Professor of Physics at the University of Turin, strada San Vincenzo 40/2C, I-10122 Turin, Italy.

What is scientific literacy?

John Durant

> The Centipede was happy quite,
> Until the toad in fun,
> Said, 'Pray which leg goes after which?'
> And worked her mind to such a pitch,
> She lay distracted in the ditch,
> Considering how to run.[1]

In the midst of the Gulf War, US President George Bush took time out to greet the American Association for the Advancement of Science, which had gathered in Washington for its annual meeting. Having pointed out that the US budget included substantial funding increases for mathematics and science education, President Bush added that, 'All sectors of society must recognize the importance of scientific literacy and strive to achieve it'.

Scientific literacy is a fashionable phrase in American and British educational circles. It stands for what the general public ought to know about science, and its widespread use reflects concern about the performance of existing educational systems. In 1987, the American professor of English Literature E.D. Hirsch Jr captured this concern with his best-selling book *Cultural Literacy: What Every American Needs to Know*.[2] Hirsch argued that the unity of American culture depended upon a common stock of generally shared knowledge, which he listed in the form of some 5000 essential concepts, dates, names and phrases covering more or less the entire world of formal learning. Prominent in this list were several hundred scientific terms, ranging from 'Absolute Zero', through 'Mutation', 'Nuclear Fission' and (amazingly) 'Ontogeny Recapitulates Phylogeny', to 'Y Chromosome'.

Over the past few years, there has been an international wave of concern about the relationship between science and the wider culture. All about us, scientists and teachers, writers and broadcasters, museum curators and science centre explainers are attempting to provide the general public with improved access to science. What, however, are all these people trying to achieve? What is meant by 'the public understanding of science' in Britain, by 'la culture scientifique' in France, and by 'scientific literacy' in the United States?

Even to ask these questions is to invite the criticism (popular in Anglo-Saxon culture, especially) that we are engaging in pointless philosophizing when there is much practical work waiting to be done. At best, it may be said, such questions are a luxury; at worst, it may be argued that they carry some risk of undermining the practical work itself. Who knows, but perhaps an exploration of underlying aims and purposes will reduce us all to the same unhappy state as the centipede, who 'lay distracted in the ditch, considering how to run'?

I believe this is a risk we have to take. In the midst of so much practical work, it is appropriate to pause and ask whether we are applying our efforts in the right way. In this article, therefore, I intend to play the part of the toad. I shall attempt to answer the question: what is scientific literacy? By this, I shall mean: what is it reasonable to hope and expect that ordinary citizens will know about science in order to equip them for life in a scientifically and technologically complex culture? My answer to these questions will suggest that much current effort at the creation of a scientifically literate culture is well-meant, but misdirected.

Three definitions of scientific literacy

It is worth distinguishing between three very different approaches to scientific literacy. All three approaches share the conviction that non-scientists living in a scientifically and technologically complex culture ought to know something about science. However, each emphasizes the importance of an entirely different aspect of science. The first puts the emphasis on the contents of science (i.e. scientific knowledge); the second stresses the importance of the processes of science (i.e. the mental and manual procedures that produce scientific knowledge, often referred to collectively as 'the scientific method'); and the third concentrates upon the social structures or institutions of science (i.e. what may be termed scientific culture). I shall discuss each of these three approaches in turn.

1. Scientific literacy as knowing a lot of science

Let us begin with what is probably the most familiar and certainly the most straightforward of the three approaches to scientific literacy. On this view, to be scientifically literate involves being well-acquainted with the contents of science; that is, it means knowing a lot of science. This, of course, is the approach to understanding science that dominates the world of formal education. Curricula are full to overflowing with the fruits of scientific inquiry – with theories and laws, with models and mechanisms – and, of course, with a vast array of facts that these interpretive schemata are intended to explain. Most students in most formal science courses that I've experienced have little time for anything beyond the mastering of the required amount of scientific knowledge.

The idea that the contents of science are the key to understanding science extends far beyond the confines of formal science education. For example, I have already mentioned Hirsch's popular book on cultural literacy.[2] In this book, Hirsch set out what he termed 'The basic information needed to thrive in the modern world'. This information spans a wide range of subjects, from sports to science. It is, if you like, a lexicon of largely taken-for-granted facts which Hirsch claims constitutes the stock-in-trade of (American) literate culture. Hirsch's promise to his readers is that, armed with this lexicon, they will be able to understand the contents of the daily newspaper, converse intelligently about current affairs, and participate meaningfully in public life.

Hirsch's standards for the culturally literate are quite exacting. Opening his lexicon more or less at random, I find the following entries: Hearst, William Randolph; heat capacity; heat of fusion; heat of vaporization; heavy water; Hector; hedonism; Heep,

Uriah; Hegel, Georg; Heisenberg uncertainty principle; Helen of Troy; helium; and hell hath no fury like a woman scorned. Opening it again a little further on, I find: Water, water everywhere, nor any drop to drink; Watson, Dr; Watson, James, and Francis Crick; watt; Watt, James (steam engine); Watts riots; wavelength; wave–particle duality; Way down upon the Swanee River; and Wayne, John. Any American citizen who possessed reasonably good recall of nearly 5000 items like this could certainly make a living playing Trivial Pursuit for money.

Cultural Literacy contains a mere list of all the things that literate Americans are supposed to know; but a year later, Hirsch and his colleagues brought out a much larger *Dictionary of Cultural Literacy* containing concise definitions of each and every item on the list.[3] Finally, physicist James Trefil went one better even than this, by teaming up with earth scientist Robert Hazen to write *Science Matters: Achieving Scientific Literacy*.[3] This book is a synoptic account of the main principles underlying all of the natural sciences – physics, chemistry, earth science and biology. Between one set of covers, Trefil and Hazen offer what they term 'the constellations of basic facts and concepts that you need to understand the scientific issues of the day'.

This is a big promise, and it is one that in my view the authors fail to keep. I have no major quarrels with their synopsis of scientific knowledge. Certainly, to my biologist's eye it seems over-dominated by the physical sciences (in particular, I find the total absence of medical science in a book on scientific literacy rather strange – after all, this is the one branch of science that has maximum interest and relevance for the general public); but in general, I admire the authors' skill in being able to chart a course through such a huge body of material with such apparent ease. No, the question is not: is their synopsis of scientific knowledge well done; rather, the question is: is a synopsis of scientific knowledge really what the public needs in order to 'understand the scientific issues of the day'? Indeed, more generally: is factual knowledge the key to scientific literacy?

I think the answer is clearly no. Knowing a lot of scientific facts is not necessarily the same as having a high level of scientific understanding. Of course, it is a good thing in itself if people can define heat capacity, heat of fusion and all the rest – I do not wish to argue in favour of ignorance; but in and of itself, such 'textbook' knowledge is not terribly illuminating. For one thing, being able to trot out a dictionary definition is not the same as actually knowing what the definition really means; and for another, even if a dictionary definition is understood it does not follow that either its place within science or its wider significance has been properly grasped.

But by far the biggest objection to this fact-oriented approach to scientific understanding is its total mismatch with the stated aim of equipping people to deal with 'the scientific issues of the day'. Overwhelmingly, the scientific issues of the day involve new knowledge, or even new knowledge in the process of being born. Frequently this new knowledge is uncertain; often, it is controversial. In other words, scientific experts may be undecided about things; and they may actually disagree with one another about matters of evidence or interpretation. In this situation, the public may be helped by a certain amount of background factual knowledge; but on its own, such knowledge is likely to be a poor guide to what is going on. For what is going on is the coming-into-being of new knowledge; and to understand this, people need to know something about the gestation or the embryology of science.

2. Scientific literacy as knowing **how** science works

The limitations of a purely knowledge-based approach to scientific literacy have been very widely recognized. For many years, science educators in Britain, the USA and elsewhere have sought to add some consideration of the nature of science to knowledge-based curricula. Trefil and Hazen's *Science Matters* devotes a mere page to the scientific method; but many science educators would now agree that this is inadequate. Instead of learning science by absorbing received wisdom, so-called 'process science' requires students to learn science by doing it; and even the British National Curriculum for Science, which is dominated by a concern to convey a minimum level of knowledge to all schoolchildren, finds space for at least some consideration of the nature of the scientific enterprise.

In fact, Hazen and Trefil's purely knowledge-based view of scientific literacy is actually rather unusual. More typical is the approach taken by the American Association for the Advancement of Science in its *Project 2061*. This project is devoted to establishing 'what understandings and habits of mind are essential for all citizens in a scientifically literate society'. In the project's first programmatic statement, a book entitled *Science for all Americans*, we find that the outline curriculum starts with a section on 'The Nature of Science' and finishes with another on 'Habits of Mind'.[5]

Even those who seek to measure objectively levels of public understanding of science have been persuaded of the need to include some estimate of how well people understand the processes of scientific inquiry. In the USA, Jon Miller is the leading survey researcher of public understanding of science. Miller has offered a three-fold definition of scientific literacy.[6] In his view, a person who is scientifically literate possesses: (a) a basic vocabulary of scientific and technical terms and concepts; (b) an understanding of the processes or methods of science for testing our models of reality; and (c) an understanding of the impact of science and technology on society. The first component in this list is roughly what Hirsch and Trefil mean by scientific literacy; but the second component is a requirement for people to understand what Miller terms 'the scientific approach'. The critical question here, according to Miller, is whether a citizen knows enough about the process of scientific investigation to be able to distinguish between science and pseudo-science.

This focus on the processes of science is to be welcomed. It is obviously desirable that the public should understand not only the key scientific principles but also the key scientific procedures by which these principles have been established. However, whereas scientific principles can be stated by any reasonably competent scientist, scientific procedures are far trickier to define. Most scientists are not taught anything very explicit about scientific processes of inquiry; rather, they tend to learn about these processes in much the same way that joiners or metal-workers learn about their respective trades – by being apprenticed to experienced, skilled people. This puts science educators who wish to say something about the nature of science in a rather difficult position.

In the absence of well-codified rules of scientific investigation, what tends to happen is that educationalists fall back upon certain informal but fairly standard images of science. Very commonly, science curricula that make efforts in the direction of teaching the processes of scientific inquiry incorporate some or all of the following elements: (i) there is a scientific approach to problem-solving; (ii) the scientific approach to problem-solving involves the adoption of the scientific attitude and the scientific method;

(iii) the scientific attitude comprises a combination of disinterested curiosity, open-mindedness, objectivity, the habit of basing judgment upon fact, etc; (iv) the scientific method involves the formulation of hypotheses and their subjection to critical test by means of suitably controlled experiments.

This list is certainly far from being an adequate representation of all the science curricula that seek to teach the nature of science. Nevertheless, I think it is a reasonable summary of some of the key ideas that are commonly to be found in them. The question is, therefore: do these ideas constitute either a true or a useful representation of the processes of scientific inquiry? On both counts, I fear that the answer is clearly: no.

First, consider the question of truth. I cannot think of anybody who has taken a serious interest in the nature of science who would be willing to subscribe unreservedly to the four propositions listed above. Putting this another way, I doubt whether there is a single natural scientist, social scientist, historian, or philosopher who has written about the processes of scientific inquiry who would agree that these processes rest upon the twin pillars of 'the scientific attitude' and 'the scientific method'. Of course, I may be wrong; perhaps there is an odd individual who holds this particular view. If so, however, then this person holds their view in the teeth of near-unanimous opposition.

Time and space do not permit me to document this claim in detail. Instead, I merely quote two anecdotal examples in support. First, here is the physicist and sociologist of science John Ziman on the subject of 'the scientific attitude':[7]

> Research scientists are supposed to acquire (or be born with) peculiar virtues of saintliness and wisdom called 'the scientific attitude', which especially befits them for leadership in the affairs of this wicked and stupid world. This nauseating doctrine ... was quite fashionable in the 1930s – until, as Robert Oppenheimer put it, the physicists had 'known sin' by making an atom bomb. It was never publicly repudiated by the scientific community, but has been sufficiently discredited by external events.

Ziman's point here is that the course of science itself, and particularly its increasing involvement in real-world industrial and military applications, has undermined the notion that scientists approach their work in a distinctive frame of mind that may be termed the scientific attitude. Of course, the scientific profession may embody the high ideals of disinterested curiosity, open-mindedness, objectivity and so on; but individual scientists, like the members of any other profession, approach their daily work with all manner of different personal attitudes. Here, what matters is the distinction between ideals or professional norms, on the one hand, and realities or professional conduct, on the other.

Turning next to the subject of 'the scientific method', it is worth noting the view of the late Sir Peter Medawar, an extremely successful scientist who also wrote interestingly and thoughtfully about the nature of science:[8]

> There is indeed no such thing as the 'scientific method'. A scientist uses a great variety of exploratory stratagems, and although a scientist has a certain address to his problems – a certain way of going about things that is more likely to bring success than the gropings of an amateur – he uses no procedure of discovery that can be logically scripted.

Medawar does not suggest that scientists use no distinctive methods in their work. On the contrary, he points out that they use a great variety of methods, or what he terms 'exploratory stratagems'. However, he insists that these methods cannot be boiled down to a formal procedure worthy of being dubbed 'the scientific method'. Thus, many scientists conduct experiments with a view to testing specific hypotheses; but equally, many do not. Significantly, the norms of science require that methods of investigation be strictly defined, and that they be clearly and explicitly stated in scientific publications (of this, more later); but science cannot be defined by the use of any single or simple method.

So much for the truth of the standard way of talking about the processes of scientific inquiry in science education. What, then, of its usefulness? At first sight, it seems obvious that an untrue description of science is unlikely to be particularly useful. Still, it is possible that a highly simplified or even grossly distorted account of the nature of science might just be of some assistance to beleaguered non-scientists. In this case, however, I doubt whether this is the case.

Consider again the original purpose of introducing discussion of the nature of science into basic science education. The aim, according to Miller, is to assist non-scientists to distinguish between science and pseudo-science. Unfortunately, however, most pseudo-scientists appear only too familiar with the standard account of the nature of science; and the first thing they do is to make quite sure that they and their work conform to it. Which examples shall we take – so-called 'creation science', with its 'young earth model' of historical geology? 'Complementary medicine', with its reflexological diagnoses and homeopathic remedies? 'New age science', with its (allegedly testable) theory of morphic resonance? 'Parapsychology', with its statistical analyses of clairvoyance and its controlled observations of spoon-bending'?

The world of pseudo-science is full of people who insist that they admire 'the scientific attitude' and that their work is carried out according to the strictest canons of the 'scientific method'. If these are the only criteria we have to go on, we are likely to have the greatest difficulty in drawing the boundary between science and pseudo-science.

3. Scientific literacy as knowing how science really works

Finally, therefore, we come to the third approach to scientific literacy. This goes beyond science as knowledge and science as idealized process to consider science as social practice. The fact is, of course, that science is an activity performed by people who belong to a professional community of scientists. This fact is so obvious that scientists and educationalists alike often pass over it as if it were of no consequence. But in reality it is of the greatest consequence. For the process of generating scientific knowledge is not something that is confined to the brains and hands of isolated individuals. Rather, it is something that necessarily extends across a network of colleagues, competitors and critics. This network is essential to the creation of new scientific knowledge; without it, all we have are bright (or stupid) ideas and intriguing (or dull) findings.

At an absolute minimum, the social process of scientific knowledge production involves: a corpus of existing knowledge; a professionally trained scientist, who has identified a 'problem' or other suitable opportunity to contribute to the corpus; the successful conduct of a piece of new work; the writing up of the work according to strict conventions; the refereeing (and possible rejection or modification) of the work; the publication of the work; the critical scrutiny of the work by an indefinite number of other professional colleagues; and (with luck) the eventual passage of the work into the corpus of existing knowledge. Science is the most impressive and successful body of accumulating knowledge that has ever been produced; it is surely no coincidence that the scientific community is also the most highly organized and efficient social system of knowledge production that has ever been invented.[9]

The most serious weakness in the standard view of the processes of scientific inquiry is its tendency to project the qualities of scientific knowledge upon the individual scientists who produce it. Scientific knowledge is (generally speaking) objective, so it is presumed that individual scientists approach their work in a spirit of objectivity; scientific knowledge is continually being revised and improved, so it is thought that individual scientists approach their work in a spirit of open-mindedness and humility; scientific knowledge is extraordinarily reliable, so it is concluded that individual scientists make use of a fool-proof method of investigation; and so on. The projection of the characteristics of science upon its practitioners is partly responsible for the public image of scientists as super-men and -women; but this projection obscures the true nature of science and makes it all the more difficult to understand the course of science in public.

Consider the way that science is commonly represented in public. Typically, new developments are described in personal terms. The drama of personal discovery attracts writers and producers because they know that personal stories are more interesting to readers and viewers. The result is often that the complex social system of knowledge production is intentionally or unintentionally distorted. Single results may be seized upon and given a significance far beyond what they really warrant; and audiences imbued with the idea that the secret to the success of science lies in the extraordinary qualities of individual scientists may be singularly ill-equipped to correct for such production bias. Here is a scientist, and he or she has discovered that such-and-such is the case; what could be simpler, or more beguiling, than that? But how often the scientist turns out to be wrong!

Consider by way of example the public debate in 1989 about 'cold fusion'. This debate began when an extraordinary scientific claim was projected internationally through a televized press conference in which two scientists described results that had not even got to the stage of an initial publication in the technical literature. The result was a brouhaha in which research groups rushed around trying to find out exactly what they should be doing to replicate the results, while lawyers filed patents, politicians organized funds, and the general public looked on in a state of bewildered perplexity.[10]

Cold fusion is a large and dramatic example of something that happens continually on a smaller scale in the reporting of science in public. Thus, for example, the results of a routine epidemiological study by a medical research team in Cardiff, South Wales, were leaked to the press because they appeared to show that the drinking of whole milk is associated with lower rates of heart disease. According to one newspaper headline, 'Butter can slice heart attack risk'. Yet at the time of this study there was a scientific consensus that the consumption of animal fat is associated with increased (rather than decreased) rates of heart disease; this single new study was uncorroborated, and in any event it made no causal claims about the relationship between animal fat consumption and the risk of heart disease; and, last but not least, a public relations company working for the dairy industry appears to have had a hand in orchestrating the publicity.

In cases such as this, publicity pre-empts the normal processes of 'quality control' that stand between an individual piece of scientific research and its adoption into the corpus of accepted scientific knowledge. In most cases, of course, science's quality control systems eventually catch up with events; sooner or later, it becomes clear whether a particular research result will stand; but in the mean time, there is scope for an indefinitely large amount of scientific, public and political confusion.

In order to make sense of high profile science, the public needs more than mere factual knowledge – of atomic structure in the case of cold fusion, or of the composition of animal fats in the case of milk consumption and heart disease; and it needs more, too, than idealistic images of 'the scientific attitude' and 'the scientific method'. What it needs, surely, is a feel for the way that the social system of science actually works to deliver what is usually reliable knowledge about the natural world. The public needs to understand that sometimes science works not because of but in spite of the individuals who are involved in the process of knowledge production and dissemination.

Conclusion

When it comes to scientific literacy, the needs of scientists and the general public are rather different. Scientists possess very detailed knowledge in the relatively restricted areas of their specialist research; beyond this, they tend to have increasingly general knowledge only. Significantly, scientists tend to be extremely critical of a lot that goes on within their restricted areas of specialist expertise. Some work is seen as excellent, but commonly a great deal is dismissed as second-rate or even worthless. Typically, scientists are quick to evaluate new claims in their own fields – judging some to be very important, others to be potentially interesting, and yet others as complete nonsense. Scientists can do this partly because they know others by reputation (reputation is the crucial currency in scientific debate), and partly because they can make their own on-the-spot quality assessments.

Precisely because they have first-hand experience in their own restricted areas of specialist expertise, scientists are quite likely to be reasonably discriminating in their approach to new findings in other fields as well. Faced with some new, apparently important claim outside their field, scientists may reflect on how well the claim fits with other, well-established findings in which they have confidence; they may ask colleagues who know more about the subject for a view; they may look to a particular well-respected journal for an opinion; or they may simply 'wait and see'. Of course, there is no guarantee that scientists will always make the right judgments in cases like these; but at least their personal research experience gives them some feel for the complex issues that are involved in the assessment of any new claim in science.

By contrast, most members of the general public have no direct experience of scientific research at all. The most they are likely to have is a limited array of predigested, 'textbook' knowledge derived from formal science education. Such knowledge is all of the 'beyond dispute' variety; that is, it is either so elementary or else so tried and tested that there is now no significant debate about it among expert scientists (or anybody else). Such knowledge is a very poor preparation for science as it is generally encountered in daily life. For public science is mostly new, and as often as not it is in process of active debate among experts who are trying to judge its quality and significance. In short, public science is generally caught up in what I have termed the quality control systems of science. In order to make sense of such science, the public needs to have a feel for the ways in which these quality control systems serve to separate the wheat from the chaff.

Formal science education has made some response to this problem by incorporating into curricula material on the nature of science. At the same time, informal science education has attempted to convey something of the spirit of scientific inquiry through, for example, hands-on exhibits that foster curiosity and the sense of discovery among children. All too often, however, these responses are limited by their dependence upon a highly idealized version of the processes of scientific inquiry; a version in which, as I have suggested, many of the characteristics of mature scientific knowledge are projected on to the personalities and practices of individual scientists. This is a positive hindrance, in that it makes more difficult the business of coming to terms with contingency, controversy and uncertainty in science. We need to consider how a truer picture of science can be conveyed to a general public which has no direct experience of scientific research at all.

References

1   Attributed to Mrs Edmund Craster, d. 1874.
2   Hirsch, E.D. Jr, 1987, *Scientific Literacy: What Every American Needs to Know* (Boston: Houghton Mifflin).
3   Hirsch, E.D. Jr., Kett, J., and Trefil, J., 1988, *The Dictionary of Cultural Literacy: What Every American Needs to Know* (Boston: Houghton Mifflin).
4   Trefil, J., and Hazen, R., 1990, *Science Matters: Achieving Scientific Literacy* (New York, London: Doubleday).
5   American Association for the Advancement of Science, 1989, Project 2061: *Science for all Americans: A Project 2061 Report on Literacy Goals in Science, Mathematics and Technology* (Washington DC: AAAS).
6   Miller, J.D., 1983, Scientific literacy: a conceptual and empirical reveiw. *Daedalus*, **112**(2), 29–48.
7   Ziman, J., 1980, *Teaching and Learning about Science and Society* (London, New York: Cambridge University Press), pp.48–49.
8   Medawar, P., 1984, *The Limits of Science* (Oxford University Press), p.51.
9   For an introduction to the social system of knowledge production, see Ziman, J., 1968, *Public Knowledge: An Essay Concerning the Social Dimensions of Science* (London: Cambridge University Press); and Barnes, B., 1985, *About Science* (Oxford: Blackwell).
10 For an account which suggests that what happened in the case of cold fusion is typical of what goes on in the process of resolving scientific controversies, see Collins, H., and Pinch, T., 1993, *The Golem: What Everyone should Know about Science and Technology* (Cambridge University Press).

Author

John Durant is Assistant Director (Head of Science Communication) at the Science Museum, London, Professor of the Public Understanding of Science at Imperial College of Science, Technology and Medicine, University of London, and Editor of *Public Understanding of Science*. He is based at the Science Museum Library, London SW7 5NH, UK.

National and international in science: a dialogue

Elisabeth Crawford

This is a dialogue between the Defender of the Faith (DOF) in international science and the sceptic (S). The DOF is probably but not necessarily a scientist; the S, probably but not necessarily a sociologist or historian of science. As shown by the references to published works, the dialogue is for the most part based on extant sources. I extend my apologies to the authors for this unconventional way of using their works.

DOF: Science is by nature universal. The truths which scientists seek to discover are not national truths; they are the same everywhere and so can be unanimously recognized. The structure as well as the nature of science is universal.... . Because of its objectivity, science is ... supracultural, untrammelled by the conflicts of values to which all other expressions of culture are dedicated; in the same way, its universality leads to the postulation of a supranational specificity for the institution which it constitutes and for its members. Even the idea of national scientific communities is contradictory; there can only be one scientific community, which must therefore be international. A single language, similar procedures, comparable experiments, shared norms – all these characteristics must distinguish scientific activity from any other.[1]

S: Before we go any further perhaps we should introduce ourselves. Who did you say you were?

DOF: I am a researcher at the CNRS.

S: Excuse my ignorance, but what is the CNRS?

DOF: CNRS stands for the Centre National de la Recherche Scientifique – it is the French national research organization. It employs about 25,000 researchers, engineers, technicians and administrative personnel and covers all fields ranging from byzantine studies to high energy physics.

S: I presume they are all in France. I am puzzled because I thought you just said that the institution of science was by necessity international. Doesn't an organization such as the CNRS show that science is predominantly national? I read recently that at least 90 per cent of today's research and development is carried out within nationally based coordinate systems, either private or public. This means that only a tiny portion, 10 per cent at most, is carried out within distinctly transnational programmes or institutions – that is those where the pooling of national resources for a common purpose creates supranational facilities or staff – and even in those, such as at CERN, for instance, national elements are still strong.[2] Where does that leave your international institution of science and your international scientific community?

DOF: But you don't understand. It doesn't matter if science is produced in national institutions because, as I just said, the scientific method, that is, the fact that knowledge claims are judged objectively according to pre-established, impersonal criteria, makes science inherently universalistic.[3] The nationality of the scientist is irrelevant. This is what Pasteur meant when he said: 'The scientist has a nation, but science has none'.

S: So what you are saying is that while costume, architecture, modes of conveyance, means of destruction, and even the design of machinery easily admit of some cultural variation, the same may not be said about the mathematical laws ascribed to the physical world. One finds $F=ma$ and $PV=nRT$, not $F=ma^2$ or $PV=nR^T$, whether taught in Rabat or Bandung, Tananrive or Quito.[4]

DOF: That's exactly it.

S: But you are referring there to the most quantitative, formalized part of the sciences, which is really an extreme example. As you move toward the qualitative parts that are much more difficult to formalize you will have to admit that there is much more influence from cultural and national contexts. But arguing about degrees of influence is not very interesting. I will reiterate my claim that the nation has been the principal supporter of modern science. National needs, both practical and spiritual, determined the evolution of scientific institutions and disciplines. But this use of science depended on the apparent neutrality of scientific publications. We think today that science speaks directly to all interested parties; works in science or scholarship must be able to 'shift for themselves' across national and linguistic frontiers, as Benjamin Franklin noted about his papers on electricity.[5] So there is really a terrific tension between the universalist ethos and national demands.

DOF: I can see that there would be national demands in wartime. Wasn't it Fritz Haber, the German physical chemist, who said: 'In wartime the scholar belongs to his nation, in peacetime to mankind'.

S: And Haber certainly lived up to this precept because he led the German chemical warfare operations during the First World War, even turning over his institute, the Kaiser-Wilhelm Institut für physikalische Chemie und Elektrochemie, to the Army – not only for research on poison gas but also for its manufacture.

DOF: I admit that during the First World War, German scientists went far out of their way to support the war effort not just by lending their expertise to the development of weaponry but also by engaging in propaganda activities. Many of the Allied scientists did the same, and after the war they in fact obstructed the resumption of international scientific relations through their boycott of German science.[6] But these were aberrations produced by abnormal conditions. When the British scientist A.J. Cock says that 'scientists should look back with some, possibly prophylactic, shame on the events of 1918 and 1919'[7] this shows that these episodes led to a lot of soul-searching in the international scientific community. And things soon returned to normal. Well, Haber even received the Nobel prize in 1918 for his pre-war work on the synthesis of ammonia from its elements. And the Nobel prize is after all the highest international honour that can be bestowed on a scientist.

S: That Haber was not censored for his gas warfare was mainly because so many prominent German academic scientists and also Allied ones took part in these activities. Had the Nobel committees required non-involvement as a condition for receiving the prize, they would have had to eliminate many valid candidates.[8]

DOF: Since we both agree that the First World War (and the Second one as well) created conditions in which the universalist ethos was put in abeyance, perhaps we can go on to other matters. What I don't understand *at all* is your statement earlier that national needs, both practical and spiritual, have determined the evolution of scientific institutions and disciplines. Do you mean to say *at all times*, that is, in peacetime too?

S: Yes, I do; but here the argument is a little bit more complicated, so let me explain it to you. It really hinges on what we mean by a nation and by nationalism. I am using the definition of Ernst Gellner: 'Nationalism is primarily a political principle, which holds that the political and the national unit should be congruent'.[9] This process of 'creating' nations is at the heart of Gellner's formulation. It is a process that is by no means restricted to the political sphere. Culture and language are the most essential elements, but not in the sense of the folk cultures and dialects encouraged by romantic nationalist movements. On the contrary, in Gellner's words: 'Nationalism is essentially the general imposition of a high culture on society, where previously low cultures had taken up the lives of the majority, and in some cases the totality, of the population. It means that generalized diffusion of a school-mediated, academy-supervised idiom, codified for the requirements of reasonably precise bureaucratic and technological communication. It is the establishment of an anonymous, impersonal society, with mutually substitutable atomized individuals, held together above all by a shared culture of this kind, in place of a previous complex structure of local groups.'[10] To Gellner, it was industrialization that created the new division of labour and the large social units that made necessary the shared high culture, which defines a 'nation'. In the industrial age, he concludes, 'such a nation/culture becomes the natural social unit ... and cannot normally survive without its own political shell, the state'.[11]

The role of scientific research in the process of industrialization is obvious and well known. But no less important than their work for the nation's material welfare was the role of science and scientists in creating the 'high culture' that made for the unity of the nation-state: upholding the tradition of the nation's great universities while at the same time making these venerable institutions a force of modernization; being *Kulturträger* – bearers of culture – infusing scientific knowledge and values into the cultural life of the nation; and generating the discoveries that would make the citizenry in general identify with and be proud of its scientists. This made for three different kinds of nationalism in science that were particularly relevant for the flourishing of the nation-state in the late nineteenth and early twentieth centuries, but, as we shall see, they also apply to the present.[12]

The first form of nationalism was the *mature*, or to use Gellner's term 'natural', *nationalism* found in nation-states of such long standing, relatively speaking, as England, France, the Netherlands and Switzerland. National scientific enterprises existed (and continue to exist) with varying degrees of autonomy. A high degree of autonomy meant that scientific practice was insulated from national needs and concerns, at least in peacetime. Scientists' attitudes toward their nations does not matter much in their scientific practice. Still, there is no doubt that in the larger scientific nations, the milieu and products most familiar to them are national ones. The United States, in particular, represents a virtually self-contained scientific universe with as much as 70 per cent of all American citations in the *Science Citation Index* referring to papers in other American journals.[13]

The second form of nationalism in science might be termed *practical nationalism* to describe a closer involvement of science and scientists with national interests and concerns. At the turn of the century, this form was found outside the European scientific centres: on the northern periphery (Denmark, Norway and Sweden) and also within North America (Canada and the United States). What defined this form of nationalism was first of all the way it was oriented toward practical actions, but also the way these involved the nation and national sentiments not only culturally but in a physical-spatial sense. Here, science linked to the management of natural resources was seen as helping to expand the inhabited part of the national territory, making the wilderness cede to 'civilization,' to use a cliché, whether it be in the American West or the Far North. During the latter part of the twentieth century, this kind of natural resources science has been practised most intensively in developing countries.[14]

The third form was the *militant nationalism* that spread throughout Europe in the late nineteenth century as 'nations' without states – Poles in Germany and Russia, Finns in Russia, Czechs in the Austro-Hungarian empire, and Welsh and Irish in the United Kingdom – sought to accede to statehood. The movements, built on nationalist sentiments, emphasized cultural and linguistic distinctiveness. Scientists participated in these movements as part of the intelligentsia, most actively when they devoted themselves to creating and maintaining the institutions of higher learning that were seen as important 'homes' for the nation. In Wales, for example, the national university set up in 1893 was the first, and for a while, the only national institution of the Welsh people. The splitting of Prague's Charles University in 1868, and Technische Hochschule in 1882, into German and Czech parts was symptomatic of the way militant nationalism had come to dominate the Habsburg multinational empire. With the recrudescence of nationalism, even hypernationalism, at the end of the twentieth century, we may see a return to the times when science was a means to assert nationhood. Does this mean we will see claims for the specificity of Ukrainian, Slovak, Croatian or Macedonian science in the years to come? It seems possible, although it is too early to tell.

DOF: You make it all seem terribly grim. Is there really no space for international science?

S: Well, this is where I have some good news for you. The internationalization of science is actually growing apace at present, but in order not to get caught in the kind of rhetoric about 'international science' you used at the beginning of our conversation, I would prefer to refer to this process as the *denationalization* of science. Also we have to be very cautious when we talk about these trends because science indicators are largely inadequate for grasping the fine structure of the present flux and reflux of scientific activities across national boundaries. But they are all we have. Take the money, for instance. At least 90 per cent of the funding for worldwide R&D comes from national sources. Still, the transnational flow of payments is growing rapidly. Between 1985 and 1989 alone the percentage of expenditure on R&D derived from foreign sources rose from 7.9 to 10.6 per cent in Canada, from 4.8 to 7.3 in France; and from 3.6 to 4.8 (1990) in Italy. We do not know how much of this increase was due to transnational spending by corporations, but in Europe at least we know that the increasingly large funds of the EC are playing a role. For the period 1990–1994 the EC research budget is set at about 1.75 billion ECU a year – approximately 2 billion US dollars. This is a considerable increase over the 1.1 billion ECU available annually during the previous framework programme 1987–1991. Still, it only represents about four per cent of public, or two per cent of public and private, research expenditure in the twelve EC countries taken together. These funds, as well as the wide array of EC programmes in the area of R&D – BRITE, COMETT, RACE, SPRINT and so on – and education – ERASMUS, LINGUA – are having a growing impact on the scientific community. Both as an effect of the programmes, and as an effect of general transnationalizing forces, multilaterally co-authored papers – that is, papers co-authored by persons from two or more countries – are growing within the EC. In most Western European countries the figures for the period 1981–1986 range between 20 and

40 per cent, whereas in the United States they were less than 10 per cent, and in the Soviet Union they only attained four per cent during the same period.[15] Figures for the late 1980s and early 1990s indicate that the co-authorship network within the EC is growing vigorously. The perhaps most interesting aspect of this new development is the positive feedback created by the fact that internationally co-authored papers receive significantly higher credit in terms of citations. Because the scientists involved seem to enjoy a higher chance of internal success in terms of citations, they can be expected to seek out possible new sources of funding for this type of communication.[16] So even if I do not want to paint too rosy a picture, and even if *science* is primarily an enterprise contained within the nation-state, when seen both as an economic and a cultural resource, we can see tendencies on many levels, primarily within regions such as the EC, towards the denationalization of *research*.

DOF: It certainly gives me pleasure to hear this. I am sorry I have to break off this very interesting discussion but I am off to Brussels for a meeting of one of the subcommittees of RACE. There is a chance that my laboratory will get funds for some experiments we want to do on integrated circuits.

S: And I am off to Paris for a meeting of Section 36 of the National Committee for Scientific Research. We are distributing funds among the research units affiliated with the Section and I have to watch out for the interests of my laboratory.

Both: As long as you're up, get me a grant.[17]

References

1   Salomon, J.J., 1971, The 'internationale' of science. *Science Studies*, **1**, 23–42; see pp.23–24.
2   Crawford, E., Shinn, T., and Sörlin, S., 1992, The nationalization and denationalization of the sciences. *Denationalizing Science: the Context of International Scientific Practice*, Sociology of the Sciences Yearbook 16, edited by E. Crawford *et al.* (Dordrecht: Kluwer), p.2.
3   Schroeder-Gudehus, B., 1990, Nationalism and internationalism. *Companion to the History of Modern Science*, edited by R.C. Olby *et al.* (London, New York: Routledge), pp.909–919.
4   Pyenson, L., 1989, Pure learning and political economy: science and European expansion in the age of imperialism. *New Trends in the History of Science*, edited by R.P.W. Visser *et al.* (Amsterdam, Atlanta GA: Rodopi), p.210.
5   Pyenson, L., 1992, review of *Nationalism and Internationalism in Science 1880–1939: Four Studies of the Nobel Population* by Elisabeth Crawford. *Times Higher Education Supplement*, October 16, p.22.
6   Schroeder-Gudehus, B., 1978, *Les scientifiques et la paix: La communauté scientifique internationale au cours des années 20* (Montréal University Press); Crawford, E., 1992, *Nationalism and Internationalism in Science, 1880–1939: Four Studies of the Nobel Population* (Cambridge University Press), pp.49–78.
7   Cock, A.J., 1983, Chauvinism and internationalism in science: The International Research Council, 1919–1926. *Notes and Records of the Royal Society of London*, **37**, 249–288.
8   Crawford, E., 1992, *Nationalism and Internationalism in Science, 1880–1939: Four Studies of the Nobel Population* (Cambridge University Press), p.74.
9   Gellner, E., 1988, *Nations and Nationalism* (London: Basil Blackwell), p.1.
10 Gellner, E., 1988, *Nations and Nationalism* (London: Basil Blackwell), p.57.
11 Gellner, E., 1988, *Nations and Nationalism* (London: Basil Blackwell), p.142–143.
12 Crawford, E., 1992, *Nationalism and Internationalism in Science, 1880–1939: Four Studies of the Nobel Population* (Cambridge University Press), pp.32–37.
13 Crawford, E., Shinn, T., and Sörlin, S., 1992, The nationalization and denationalization of the sciences. *Denationalizing Science: the Context of International Scientific Practice*, Sociology of the Sciences Yearbook 16, edited by E. Crawford *et al.* (Dordrecht: Kluwer), p.4.

14   Adhikari, K., 1991, Producing knowledge about natural resources: the case of scientific research on rice in India. *Social Science Information/Information sur les sciences sociales*, **30**, 445–470.

15   Crawford, E., Shinn, T., and Sörlin, S., 1992, The nationalization and denationalization of the sciences. *Denationalizing Science: the Context of International Scientific Practice*, Sociology of the Sciences Yearbook 16, edited by E. Crawford *et al.* (Dordrecht: Kluwer), pp.5–6.

16   Leydesdorff, L., 1992, The impact of EC science policies on the transnational publication system. *Technology Analysis & Strategic Management*, **4**, 294.

17   This is the title of an undated guide to proposal writing techniques by Harold A. Wooster of the US Air Force Office of Scientific Research.

Author

Elisabeth Crawford is a senior research fellow of the Centre National de la Recherche Scientifique and affiliated with the Groupe d'Etude et de Recherche sur la Science de l'Université Louis Pasteur (GERSULP), rue Blaise Pascal, 67070 Strasbourg, France.

Programming nature and public participation in decision making: a European perspective

Peter E. Glasner

Introduction

The initiative to map the human genome is a worldwide research effort costing billions of dollars. It aims to analyse the structure of human DNA and to determine the location of the estimated 100,000 human genes. The total sequence is about 2.8 billion linear bases on 23 chromosomes, of which only about 40 million are currently mapped. It has been described as the biological science's equivalent of putting people on the moon. The information collected is expected to be the source-book for biomedical science in the twenty-first century, and will be of immense benefit to the field of medicine. Most of the major single gene disorders and some of the genes involved in complex diseases should be known within the decade. The basic data will be collected in electronic databases that will make the information readily accessible in a convenient form to all who need it. These developments have major social implications.[1]

The human genome mapping programme is an international initiative. It began in 1988 with the United States, Britain, Japan and the EC as its key participants, and is co-ordinated in Europe by the Human Genome Organization (HUGO), whose headquarters are currently in London. HUGO's brief is to provide a forum and intellectual leadership for international scientific debate at the highest level. It is required to encourage public debate on the scientific, ethical, legal and commercial implications of human genome projects. However, it has shown some hesitation in fulfilling this role.[2]

There are a number of key issues related to the human genome programme which should be addressed in debates aimed at educating the lay public and at involving non-experts in the decision-making processes, and which do not involve addressing the more specialist questions of the nature of molecular biology. These include most centrally:

(a) Fairness in the use of genetic information, for example with regard to employment, the criminal justice system, etc.
(b) The impact of knowledge on the individual, for example stigmatization and labelling.
(c) Privacy and confidentiality.
(d) The impact on genetic counselling.
(e) The impact on reproductive decisions.
(f) Issues for medical practice, such as education, standards of quality control, etc.
(g) The relevance of past uses and misuses of genetics.
(h) Questions raised by commercialization, such as intellectual and other property rights.[3]

There is growing public awareness of and debate about these implications among non-experts in the USA and in some European countries. But progress is slow, and it is legitimate to ask why. The various agencies associated with funding the human genome programme have all set aside a small proportion of their funds in order to address these issues. However, the debate is not regarded with the same degree of concern by the public as the momentous breakthrough achieved by discovery of how to engineer genes, even though its impact is likely to be at least as far-reaching.[4]

HUGO: the 'United Nations' for the human genome

The Human Genome Organization was conceived in April 1988 at a genome-mapping symposium at Cold Spring Harbor in the USA. The acronym HUGO was suggested by Sydney Brenner, a molecular biologist from Cambridge University. It was born in Montreux, Switzerland, in October of the same year, with funding from various charities, and a Council was established with 42 members – including five Nobel Laureates – representing 13 nations. By 1991 its worldwide membership, open to all persons concerned with the human genome and related scientific subjects, had reached 240 and was still growing. Its Articles set out the purposes of HUGO as follows:

> to assist with the co-ordination of research on the human genome and foster co-operation between scientists to avoid unnecessary competition or duplication
> to co-ordinate and facilitate the exchange of relevant data and bio-materials
> to encourage public debate, and provide information and advice on the scientific, ethical, social, legal and commercial implications of human genome projects

This last objective was to be implemented through one Committee on Ethical, Social and Legal Issues, and another on Intellectual Property and Ownership. The European headquarters were established in London under the Directorship of Sir Walter Bodmer and with funding from the Wellcome Trust.

Rather like the United Nations, to which it has been likened by the American zoologist Norton Zinder, HUGO was not universally welcomed; nor has it been entirely successful in pursuing its objectives. It was considered by many scientists at its inception to have too circumscribed a membership. Daniel Kevles notes that in France, in particular, there was a built-in resistance to its scale and degree of centralization from scientists socialized into a culture derived from the Pasteur Institute model and focused on small-scale, artisan modes of research.[5]

> To many Pasteurians, human genome research threatened to put a premium on managerial and technological skills, smother Little Science, and take away its resources.

Even in 1988, Robert A. Weinberg observed that important decisions about the future development of the human genome programme had in effect already been made, possibly to the detriment of more conventional research in molecular biology and certainly with no reference to an international body like HUGO.[6] The European Science Foundation, in its 1991 Report on Genome Research, suggested that HUGO:[7]

... will not, and indeed cannot represent an ideal mechanism for co-ordination of the European genome effort. Europeans do and increasingly will contribute to the membership of HUGO, but specific European problems will not be resolved through a body with such a broad remit.

In 1990 at a closed session of a UNESCO-sponsored conference on the human genome project in Paris, a German scientist called upon HUGO to publish a declaration affirming that the carrier of a genome has the exclusive right to knowledge of his or her genotype. The president of HUGO, Sir Walter Bodmer, stated that while HUGO would identify ethical issues for consideration, it could not tell people of different countries and cultures what to do.[8]

The HUGO Committee on Ethical, Social and Legal Issues met for the first time in October 1992 in order to establish a framework and agenda for future work. It expects to meet once again late in 1993. While the major funder of the HUGO office, the Wellcome Trust, is well disposed towards the work of the Committee, it will be up to the members, having identified projects, to seek sources of funding from elsewhere, for example from the EC. It is therefore likely that the Committee will continue to move slowly in this area, and HUGO will not be seen to contribute significantly to the on-going debates for some time to come. Since HUGO does not publish either an Annual Report or a full list of its members, it is difficult to be precise about the exact nature of its evolving agenda in the area of ethical, social and legal issues.

European public reactions to the human genome programme

The new genetic technologies differ somewhat from other forms of technology in that social debate and conflict often seems to precede rather than follow their introduction. Public attitudes are likely to be influenced by a number of factors such as cultural values, nationality, socio-demographic variables, information (or the lack of it), and distrust both of the promoters and of the estimates of potential risks. Two recent studies, *Eurobarometer 35.1* and *Bioethics in Europe*[9] throw some light on the topics which many Europeans feel are important in the human genome programme, and which should form the basis for discussion by the HUGO Committee on Ethical, Social and Legal Issues.

The Spring 1991 Eurobarometer opinion poll, one of a number undertaken twice a year since 1973, included 13 questions on biotechnology. These questions were fielded in order to achieve a better understanding of the views of Europeans in the 12 member states of the EC. A major finding was that a large number of people, particularly in Greece and Spain and especially in Portugal, were unable or unwilling to answer a number of the questions. The conclusions of the bioethics report echoed this finding, expressing the view that

> ... for the most part, the Northern European countries tend to be more active in and informed about the new biological and genetic technologies.

While, according to Eurobarometer, the two principal sources of information for Europeans were the television and the newspapers, the most reliable were thought to be consumer and environmental organizations, and schools or universities. According to the bioethics report, concerns about the negative consequences of the human genome programme were most strongly raised in Denmark, Germany, and to some extent the United Kingdom. This was attributed mainly to mass information campaigns (in Denmark) and several publicly and privately sponsored reports (in Germany), as well as the active participation of public interest groups representing sufferers from genetically inherited disabilities. Public opinion in France, Greece, Italy and Spain, on the other hand, was shaped more by the press, which stressed the potential benefits of the human genome programme and de-emphasized the ethical implications.

One European in two, however, believes that the new genetic technologies will improve our way of life over the next 20 years. The best informed groups – men, young people, the better educated and the more comfortably off – are also, according to Eurobarometer, the most optimistic. But, regardless of nationality, the large majority of respondents considered that these developments needed to be controlled by the government. Only one in ten feels that our way of life may be in danger. Bernard Dixon notes that:[10]

> With the exception of France, Italy, and Spain ... virtually all European countries have organizations actively opposing some applications of biotechnology.

In general, Dixon concludes, these public interest groups have been most effective when associated with political movements such as the Green Party. The debate has probably been most complex in Germany, with its continuing reappraisal of the eugenicist policies of the Nazis. The current situation, however, suggests that the tone may be changing: the fundamental rejection of any form of genetic manipulation is being replaced by a willingness to talk to those who are using the techniques safely.

Andreassen Rix views this change as a by-product of the decision to drop the idea of 'predictive medicine' as the focus for the modified European genome programme, 'Human Genome Analysis', which was launched at the end of 1989 and adopted in 1990.[11] It now addresses the problems associated with the location of genes of medical importance through the improvement of the genetic map, rather than gene therapy in germ cells, or intervention in the development of embryos which might lead to hereditary changes. As noted earlier, this change is likely to be linked to the disproportionate representation of Green Party members on the European parliamentary committees which most often deal with biotechnology. It has had very little to do, in any obvious way, with the direct intervention of the Human Genome Organization.

Contextualizing public participation

This discussion attempts to link four related but distinct conclusions from research undertaken in part by the author[12] in order to address the issue of why the 'moon shot' of modern biology with its far-reaching consequences has yet to surface as an issue of genuine public debate in all European countries. These four conclusions are concerned with:

1. The importance of contextualizing scientific and technological knowledge in the process of informing the non-expert about its wider issues;
2. The need to involve the lay public in exploring these issues in an informed way so that appropriate decisions about the future can be made;
3. The need to recognize that scientific expertise is to an extent socially constructed, and dependent upon particular historical circumstances; and
4. The need to explore the policy issues involved in recognizing the importance not only of the key actors but also of the social and cultural milieu in which their negotiations occur.

I suggest that together these provide a useful perspective in helping to explain why significant public debate about the implications of the human genome programme varies so greatly across Europe. There was, as noted earlier, concern enough when molecular biologists and other scientists called a halt to experiments with recombinant DNA because of fears for the effects unsupervised research might have on the future of the human race. Clearly the very fact that a scientific moratorium was successfully called added legitimacy to public fears about the potential dangers posed by genetic engineering.

However, the moratorium did not occur in a cultural vacuum. Dangers selected for public concern differ according to the strength and direction of social criticism. The perceived hazards associated with, for example, the cancer-causing properties of acquiring a leisure-time sunburn are very different from those resulting from asbestos poisoning, even though mortality rates may be similar. The attachment of moral blame after all reflects the wider cultural perceptions of society, as Douglas and Wildavsky point out:[13]

> No doubt water in fourteenth century Europe was a persistent health hazard, but a cultural theory of perception would point out that it became a public preoccupation only when it seemed plausible to accuse Jews of poisoning wells.

The development of genetic engineering techniques followed closely on real public concerns about the negative impact of some key technological advances, triggered by Rachel Carson's *The Silent Spring* and characterized by Maddox as 'the doomsday syndrome'. Current concerns about the environment, however, do not appear to have the same general impact on public perceptions in all European countries.

Research on the public understanding of science suggests that a much more localized view prevails which relates environmentalism to more salient issues, as illustrated by the phenomenon of NIMBY – 'not in my back yard'. Thus it comes as less of a surprise that concerns about the human genome project have surfaced in public debate in such countries as Germany, where the level of salience of issues concerning the eugenicist ideology which may underpin it reflect worries about a resurgence of fascism.[14]

The cultural perception theory is not, however, enough to fully explain the problem, even though we may accept that the United States, for example, is culturally more likely to view scientific and technological advance more critically than, say, Great Britain. It would appear, at least, that the American public is more interested and better informed about such matters than their British and European counterparts, particularly where medical discoveries are concerned. In general, Americans perceive science as more central to their daily lives. However, when asked about risk factors concerning health, Americans were 'more inclined than the British to see *all* risks as contributory factors'.[15] The Eurobarometer poll concludes that, while the public in Denmark have the strongest perception and the public in Italy the weakest, the perception of risk across the member states as a whole, associated with different applications of biotechnology/genetic engineering, falls within a fairly small range, and is not very high.

Other contextual factors are also significant in explaining the different levels of public participation in the debate over the development of the human genome mapping programme. For example, those biologists not directly involved in the research are not all in agreement that such a substantial proportion of the science budget should be devoted to pursuing a 'holy grail'. On an international comparison, the USA is currently spending $164 million, Japan $20 million, France $5.3 million, and the UK $4.5 million per annum on genome research. Scientists raise questions about the high 'dross' element in charting the make-up of the genome when we know so very little about the functions of all but a very small proportion of it. They fear for other projects which may not appear to have much more immediate medical benefits, but which remain central to the development of the biological sciences. They note with concern the significance of the commercial element associated with attempts to patent certain sequences, and are suspicious of the motives of their fellow researchers.[16]

These elements have entered the public debate in America. They have not done so Europe to the same extent simply because the approach here has not been to spend vast sums of public money on the human genome programme, but to focus on the 5 per cent of the human genome that actually represents coding sequences or genes (complementary or cDNA). Other laboratories have concentrated on homologous genes from non-human sources such as mice and worms. Nor has any attempt yet been made to take out patents for private gain. Hence the opening for lay participation through disagreement between experts is nothing like as clear in Europe as it is in the United States.

Neither the media nor the public in Europe paid much attention to the patenting issue until the announcement by the American National Institute of Health to allow a request by one of its researchers to patent the intellectual property of more than 2000 human DNA sequences. They were derived from participation in the human genome programme and their functions are still unknown. HUGO has consistently argued against patenting on the grounds that success in the programme depends upon a free flow of information, which is the very antithesis of asserting intellectual property rights. For the public at large, according the bioethics report, this objection is overlaid by an ethical opposition to the possibility of patenting human genetic material at all.

The commercialization of the human genome programme has also had an impact on its leadership, and in a most public fashion. James Watson, joint winner of the Nobel prize for the discovery of the structure of DNA, and the doyen of American biologists, has had to resign from his directorship of the American programme because of allegations about his family's financial interests in biotechnology and in sequencing companies with a keen interest in exploiting any results. However, this is apparently a smokescreen to hide the key source of his disagreement with Bernadine Healy, the Director of the National Institute of Health. She is in favour, and he vehemently against, the principle of patenting cDNA sequences. These events are currently seen as posing a potential threat to future funding for the human genome programme, no doubt in part because they reveal a dispute between experts which allows the entry of lay players into the drama. But in Britain, for example, no real public debate on the ethics of patenting human genetic material has yet developed. The battle between Healy and Watson is presented for popular consumption more as a battle of the sexes than a dispute between experts.[17]

In conclusion, it is clear that the European community is by no means united in its approaches to the new biological technologies, and that the nature and level of public participation and understanding varies significantly from country to country. It is evident that explanations for the unevenness of public debate about the social implications of the human genome programme in the member states of the European Community are complex. It is also the case that while funding may become available within organizations such as the Human Genome Organization for raising such a debate, apparently little has yet happened. Insights from research involving similar developments in other areas of science and technology appear to suggest a number of fruitful avenues of exploration for future research.

References

1   The background to the development of the programme as a whole can be found in: Davis, J., 1990, *Mapping our Code: The Human Genome Project and the Choices of Modern Science* (New York: Wiley); Bishop, J., and Waldholz, M., 1990, *Genome* (New York: Simon and Shuster); Kevles, D.J., and Hood, L., 1992, *The Code of Codes: Scientific and Social Issues in the Human Genome Project* (Cambridge, MA: Harvard University Press). A more technical account can be found in: US Office of Technology Assessment, 1988, *Mapping Our Genes – The Genome Projects: How Big? How Fast?* (Washington, DC: Office of Technology Assessment).

2   This was reported in Brown, P., and Concar, D., 1991, Where does the genome project go from here? *New Scientist*, 17 August, pp.13–14. See also: European Science Foundation, 1991, *Report on Genome Research* (Strasbourg: ESF; Medical Research Council, 1991, *The UK Human Mapping Project, Project Manager's Report* (London: Medical Research Council).

3   Based on Appendix 7 of *Understanding Our Genetic Inheritance. The US Human Genome Project* (Washington, DC: US Depts of Health and Energy).

4   Krimsky, S., 1982, *Genetic Alchemy: the Social History of the Recombinant DNA Controversy* (Cambridge, MA: MIT Press).

5   Kevles, D.J., and Hood, L., 1992, *The Code of Codes: Scientific and Social Issues in the Human Genome Project* (Cambridge, MA: Harvard University Press), p.29.

6   Bishop, J., and Waldholz, M., 1990, *Genome* (New York: Simon and Shuster), p.224.

7   European Science Foundation, 1991, *Report on Genome Research* (Strasbourg: ESF).

8   Turney, J., 1990, Human genetics project raises spectre of Nazi war crimes. *Times Higher Education Supplement*, February, p.16.

9   Marlier, E., 1992, Eurobarometer 35.1: opinions of Europeans on Biotechnology in 1991. *Biotechnology in Public: a Review of Recent Research*, edited by J. Durant (London: Science Museum), pp.52–108; Terragni, F., 1992, *Bioethics in Europe: the Final Report* (Luxembourg: European Parliament, Directorate General for Research).

10   Dixon, B., 1993, Who's who in European antibiotech. *Bio/Technology*, 11 January, pp.44–48.

11   Rix, B.A., 1991, Should ethical concerns regulate science? The European experience with the Human Genome Project. *Bioethics*, 5(3), 250–256.

12   In genetic engineering: Bennett, D., Glasner, P., and Travis, D., 1986, *The Politics of Uncertainty. Regulating Recombinant DNA Research in Britain* (London: Routledge & Kegan Paul) In health hazards: Glasner, P., and Travis, D., 1990, Unhealthy displays? Trade Unions, VDUs and the social construction of a health hazard. *Deciphering Science and Technology. The Social Relations of Expertise*, edited by I. Varcoe *et al.* (London: Macmillan). In public understanding of science:

Rothman, H., Glasner, P., and Adams, C., 1993, Plants, proteins and currents: rediscovering science in Britain. *Misunderstanding Science*, edited by A. Irwin and B. Wynne (Cambridge University Press), and other contributions to this book.

13   Douglas, M., and Wildavsky, A., 1982, *Risk and Culture* (San Diego: University of California Press), p.7.

14   Kevles, D.J., and Hood, L., 1992, Out of eugenics: the historical politics of the human genome. *The Code of Codes: Scientific and Social Issues in the Human Genome Project* (Cambridge, MA: Harvard University Press).

15   Evans, G., and Durant, J., 1989, Understanding of science in Britain and America. *British Social Attitudes: Special International Report*, edited by R. Jowell *et al.* (Aldershot: Gower).

16   Davis, J., 1990, *Mapping our Code: The Human Genome Project and the Choices of Modern Science* (New York: Wiley), Chap. 5.

17   Anderson, C., 1992, US genome head faces charges of conflict. *Nature*, **356**, 463; Roberts, L., 1992, Why Watson quit as project head. *Science*, **256**, 301–2; Cornwell, J., 1992, Gene spleen. *Sunday Times Magazine*, 26 July.

Author

Peter Glasner is Professor of Sociology and Dean of the Faculty of Economics and Social Science at the University of the West of England, Frenchay, Bristol BS16 1QY, UK. His books include studies of the regulation of biotechnology in Britain, and the risks and hazards associated with computer-based technology. He is currently involved in a project on the social aspects of the human genome mapping programme.

The public image of CERN

John Krige

In 1983 CERN scientists detected the W and Z vector bosons, and made reasonably precise estimates of their masses. This was a major discovery: indeed, the Nobel prize for physics for 1984 was awarded jointly to Carlo Rubbia, the spokesman of one of the experiments that discovered these particles, and to Simon van der Meer, a CERN accelerator physicist. It was also a major political success. It fulfilled the goals laid down for the laboratory by the founders of CERN: to restore the field of particle physics to its pre-war eminence in Europe, and to compete on an equal footing with the United States. Finally, the achievements were turned into a media event. The discovery of the W was announced at a press conference held at CERN on 25 January 1983, less than a fortnight after the first preliminary results were presented at a physics conference in Rome. The detection of the Z was also the subject of a CERN press release. Indeed, it is striking that the laboratory management's desire to 'market' these findings was so great that they were reported in the mass media before they were published in the scientific literature. The public at large were presented with apparently well-confirmed conclusions before the results had passed through the usual mechanisms of peer review.[1]

This rapid exposure of two of CERN's major scientific findings to the full glare of publicity in 1983 was simply one initiative among many which the laboratory has undertaken to explain its work and achievements and to improve its visibility. During the last five years in particular, the management has made increasingly extensive efforts to communicate better with the public both through television and the press and through more direct contact. A few examples will serve to give the flavour of its activities.

In November 1989, CERN inaugurated its new LEP machine, an electron–positron collider built in a huge underground tunnel, 27 kilometres in circumference, stretching from Geneva airport to the Jura mountains. A special press day was arranged in June in anticipation of the event, and was attended by about 100 journalists. A huge tent was erected for the sumptuous opening ceremony itself, which was attended by some 1500 distinguished guests who heard speeches from, among others, the King of Sweden and the President of France. The official proceedings were broadcast live on Swiss television. Extensive coverage worldwide was ensured by the presence of over 200 media representatives, including 18 television crews.[2]

The LEP inauguration ceremony was a typical upmarket public relations exercise. Its aim was to make the man and the woman in the street aware of the importance of the laboratory using predominantly political rather than scientific images. CERN's success as an example of European collaboration was enhanced by the prestige and glamour of the guests of honour and shared not only with those who feasted at its tables but with the 'hundreds of millions of viewers' all over the world who were hungry for sensational events.

Happenings like this occur against the backdrop of an ongoing effort by CERN to educate the public about the content and significance of its physics and its facilities. In 1989 alone 22,000 visitors swarmed across the site where they could walk through a newly inaugurated exhibition spread over 600 square metres, watch films explaining what the laboratory does, and have guided tours of some of its work places. Determined to inform the public in even more imaginative ways, the laboratory management also organized a day of science at Expo 92 in Seville. The experimental demonstration of scientific principles was combined with a parade of scientists and gymnasts dressed as particles, printed circuits, atoms and so forth dancing around a particle accelerator. Science, as the *CERN Bulletin* put it, was brought 'out of the laboratory and into the street'.[3]

CERN is trying to forge closer links not only with the general public, but also with industry. In 1985 it set up a special committee to explore the possibilities for increased collaboration with European industry and to export technology out of laboratory and into the world of business. This was followed by the setting up of a number of expert technical boards in various fields of engineering and electronics, and later by the establishment of an 'Industry and Technical Liaison Office'. In short, CERN is now making a concerted effort to be seen not simply as a scientific and political success, but also as a stimulus to advanced technological development in Europe.[4]

Two comments are apposite at this point. Firstly, it is of course true that CERN has for many years maintained relations with the press and the public at large, including industry. What is novel is the growing intensity and scope of its activities in the public relations field, and its determination to seek every possible opportunity to enhance its impact and to sharpen its profile. This determination was backed by the appointment in the Spring of 1989 of a new and very dynamic head of its media section and, more recently, by the creation of a high-level post to deal specifically with 'Communications and Public Education'.

Secondly, it is striking that when speaking about its science, CERN concentrates on its physics and the equipment needed to do it. To its credit, the laboratory makes a very real effort to translate abstract and extremely complex ideas into relatively simple language and everyday imagery. At the same time, no visitor to CERN will come away knowing how much the laboratory costs, how the money is spent, what interest groups shape its trajectory, or what battles the management has with the member states' delegates. In sum, apart from routine gestures towards its 'founding fathers' and to its international character, the image that CERN projects of itself is narrowly 'internalist'. The laboratory is abstracted from the social, political, economic and institutional context in which it is located and presented simply as doing 'pure' science which has increasingly important technical spin-offs.

In a certain sense there is nothing surprising about this. It is a symptom of the rigid division which physicists regularly make between scientific activity and the 'non-scientific' context which supports that activity. It is of course useful and meaningful to draw this distinction for certain purposes. It is somewhat paradoxical that CERN should persist in doing so in its public relations exercises. No-one today doubts the excellence of the science and engineering done at CERN, which is universally recognized as an outstanding high-energy physics laboratory. What is being questioned by governments is not the content of the science, but its cost. In short and too simply, if there is one single factor behind CERN's drive to sell itself better in the member states, that factor is money.

In the twenty years between 1960 and 1980 CERN's annual budget rose steadily from about 200 million Swiss francs (MSF) in 1991 prices, to 900MSF. For the past decade, however, the laboratory has been forced to peg its expenditure at the 1980 level. This has had to be done despite a doubling in the number of visiting scientists at CERN, and despite the demand made by the member states in 1981 that, exceptionally, if CERN wanted the LEP, it would have to construct it within constant annual budgets. To meet this constraint the laboratory management was obliged to make certain cuts, the most drastic of which was to close down one of the existing machines, the Intersecting Storage Rings. It was also forced to let the construction time of the new accelerator slip by one year, and to approach its design energy in stages. In a first phase, the so-called stripped-down LEP, 50GeV beams of electron and positrons would be used. In a second stage the beam energies would be increased to 100GeV (the so-called LEP-200 version) by the addition of superconducting radiofrequency accelerating cavities.

The stripped-down LEP was commissioned in 1989, but in a climate of growing financial difficulty. In undertaking to build the machine within the ceilings demanded by governments, the management made 'no allowance for contingencies and postponed payment for unforeseen difficulties until after 1988'.[5] The inevitable cost overruns, though relatively small, and the demands by creditors that they be paid, placed enormous strains on the CERN budget in the late 1980s. At the same time, for both technical and financial reasons, the upgrading of the stripped-down LEP has had to be approached in stages, and there are now doubts that it will reach its full energy of 100GeV per beam by 1994, as was earlier hoped.

Finding the money to bring LEP up to its design energy has not been the only, nor perhaps even the most important, financial headache the CERN management has had over the last two or three years. It has also had to fight for resources for the next generation machine, the Large Hadron (i.e. essentially proton-proton) Collider, or LHC. This will be the most complex and difficult machine ever built at CERN. It is to be installed in the LEP tunnel above the existing collider, and is estimated to cost about 2000MSF in 1991 prices. To bring it to fruition will require major research and development efforts both on the accelerator itself and on the detectors that will be needed to probe the collision processes. Again, member states have made it clear to the CERN management that the LHC, if built, must be done within a constant budget level. To meet this requirement CERN has to find 600MSF of new money to ensure project authorization before the end of 1993. The management hopes that 200MSF of this will be raised by 'voluntary contributions from the Member States [...]'. And in a step unprecedented in CERN's history it hopes that the remaining 400MSF of the cost of the machine will be borne by non-member states, such as Canada, China, India, Japan and the Soviet Union.[6]

CERN's major public relations efforts have to be seen against this backdrop. By alerting industry and the public at large to its existence, achievements and opportunities it hopes to mobilize support for its activities, and to make it politically difficult for governments not to fund its programmes. It also hopes to stimulate a renewed interest in physics among university students, where the field is not nearly as popular as it used to be. At the same time of course, it has to be said that if governments are reluctant to support growing budgets it is not simply a question of money. It is, more fundamentally, a question of priorities. And unluckily for CERN, the present historical conjuncture is not one conducive to major expenditure in high-energy physics, be it in Geneva or indeed anywhere in the world. This is partly due to the economic crises and the loss of confidence in the political elites that characterize much of the economically developed West. Relevant, too, is the cost to Germany of reunification: Germany is the prime contributor to the CERN budget, and bears over 20 per cent of the laboratory's financial burden. But the problem is also due to a number of considerations specific to the discipline itself.

Firstly, high-energy physics has to a certain extent become a victim of its own success. During the past two decades work on both sides of the Atlantic has confirmed to a remarkable degree the validity of the so-called Standard Model. This is a collective name for two theories: one developed by Glashow, Salam and Weinberg which unifies the weak and electromagnetic forces in the nucleus into a single electro-weak scheme, and the other – quantum chromodynamics (QCD) – which describes the strong force. No known data contradict the Standard Model, which has been confirmed by CERN's LEP to an accuracy of better than 1 per cent.[7] And although physicists insist, probably rightly, that it is not the last word, and that there are interesting discoveries to be made at accelerator energies well beyond those now available, one can well imagine governments wondering whether the time has not come to be satisfied with what we have got. This is so particularly when knowing more costs so much, and can (depending on the source of funds) be at the expense of other basic science fields which have been relatively starved of resources to feed the seemingly insatiable appetite of high-energy physics.

CERN is also suffering from the determination of governments to be increasingly selective about the basic research that they fund. For the first two decades after the war, governments steadily increased their investments in R&D, notably on the military and the nuclear fields, believing more or less implicitly that basic research automatically led to technological development and economic growth. High-energy physics was one of the beneficiaries of this cornucopia. Indeed, all of the high-energy physics laboratories which are now operational were authorized in this period. During the 1970s, by contrast, in the face of a general economic slow-down and the oil crisis, R&D expenditures as a percentage of GNP tended to drop back in most countries, with the notable exceptions of Japan and, to a lesser extent, West Germany. The figures recovered in the 1980s, but along with a new determination to couple public expenditure on basic research to specific technologies and industries. Reagan's Strategic Defence Initiative programme (SDI or Star Wars) and the European Eureka and ESPRIT programmes were typical of this new mood. CERN was ill-placed to benefit from this strategy. Indeed, a special committee set up to review the functioning of the laboratory in the mid-1980s concluded that 'the overall value of contracts considered as "technically interesting" negotiated by CERN with firms in the 14 Member States does not exceed, on average, 60 Million Swiss Francs per annum, which is marginal in terms of European industry as a whole [...].'[8] It is in this light that we should understand the increasing emphasis which CERN has placed since the mid-1980s on the technological benefits flowing from the laboratory, and the measures it has taken to forge new and productive links with industry.

Finally, the end of the cold war and the collapse of superpower rivalry has forced a fundamental rethinking of national scientific and technological priorities in the West. Until a few years ago high-energy physics in the United States could count on support on grounds of national prestige – the stream of successes in the field were one indicator of American scientific and technological superiority – and of military relevance – most explicitly, in the particle beam weapons foreseen in the Star Wars armoury. European high-energy physicists benefited from this. Each new, more powerful accelerator in the

USA generated the demand for an equivalent or better machine on this side of the Atlantic. The need to remain competitive and to avert the 'brain drain' of a strategic elite more or less ensured CERN's continued expansion. Times, and priorities, have changed. The old 'political' arguments for building bigger and more powerful accelerators are wearing thin.

In conclusion, it is becoming more and more difficult to justify to governments the increasingly large expenditures for high-energy physics research, as the CERN management knows only too well. There is also, as I have said, a growing lack of interest in the field among high-school students and university graduates. Building the public image of CERN as a scientific, technological and political success is one response to the ensuing structural crisis confronting the laboratory. It is a campaign which will undoubtedly help to raise the scientific literacy of the public. It will also inspire those who despair at the slow and tortuous path towards European integration. It remains to be seen, though, whether it will succeed in its more ambitious aim of securing a healthy financial future for the laboratory.

References

1   The paper announcing the discovery of the W was received by the editors of *Physics Letters* on 23 January 1983, the press conference was held on 25 January 1983, and the paper was published in the edition of the journal dated 24 February 1983. The paper announcing the discovery of the Z was received on 6 June 1983 by the editors of *Physics Letters*, the press release was made several days earlier on 1 June 1983, and the paper was published in the edition of the journal dated 7 July 1982. There is a separate article to be written about the increasing tendency of scientists to go public with important discoveries before they have appeared in the scientific literature, so bypassing the traditional peer review system.
2   CERN, 1989, *Annual Report*, Vol. 1, pp.34–35.
3   CERN, 1992, *CERN Bulletin 42/92*, 12 October.
4   For this paragraph see CERN, 1987, *Annual Report*, pp.37–45.
5   CERN, 1985, *Annual Report*, p.16.
6   See document CERN/CC/1891, 25/11/91, *Guidelines for the Large Hadron Collider (LHC)*.
7   CERN, 1991, *Annual Report*, Vol. 1, p.10.
8   See CERN Review Committee, Final Report CERN/1675, 3/12/87, p.34.

Author

John Krige has worked and published extensively on the history of CERN, and is currently leading a project to write the history of the European Space Agency. He is a (part-time) professor in the Department of History and Civilization at the European University Institute, San Domenico di Fiesole, Florence, Italy.

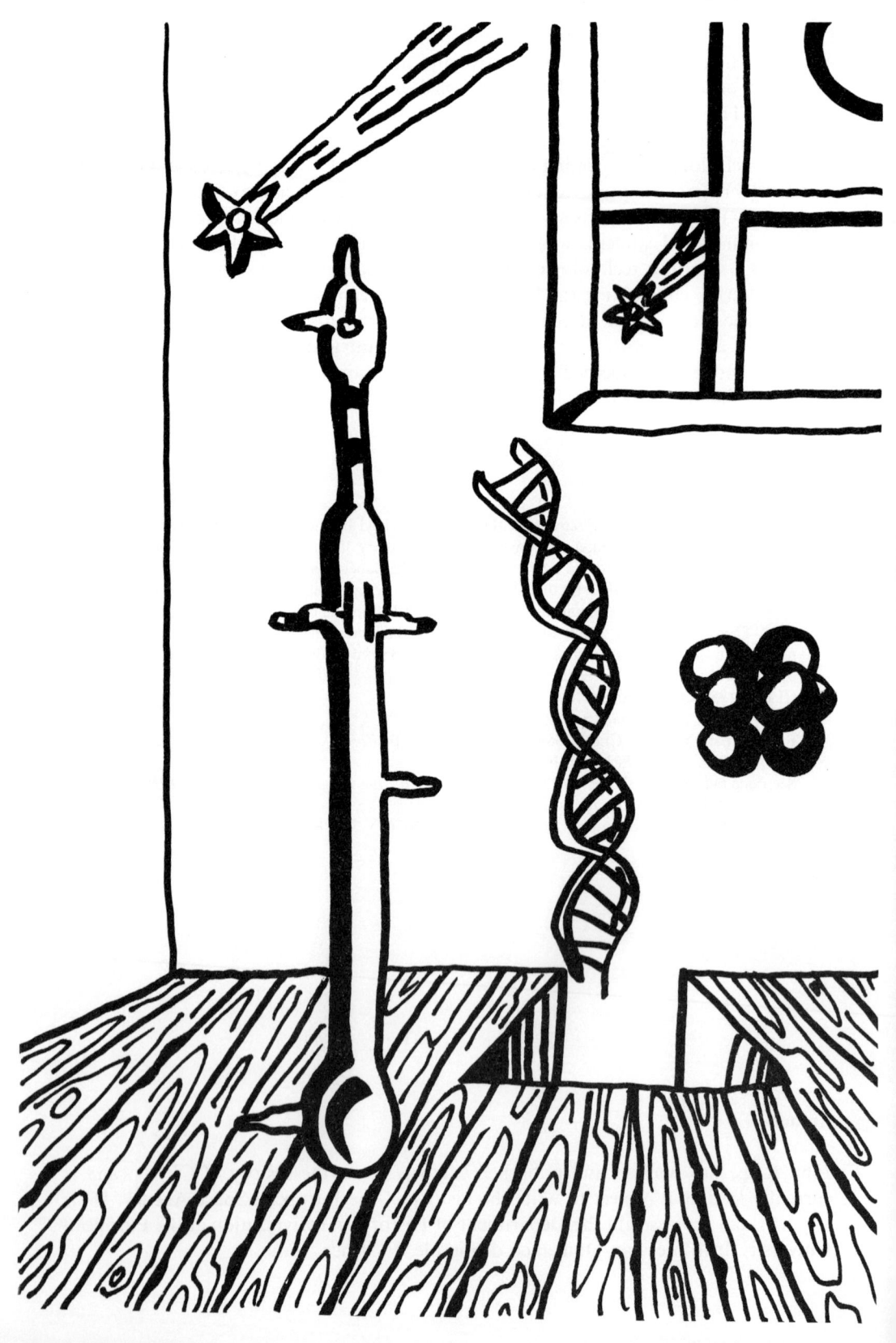

Media

The Hypothesis project

Claudio Carlone and Maria Vitale

A vision of the scientific world as the province of the few – a separate culture with an autonomous language and subject matter – still prevails among those responsible for science in Italy and dominates the outlook of the academic community, which bases its power on the logic of partition rather than the more collegial criterion of peer review. The absence of a tradition of science journalism and the indifference of those in charge of the media has given rise to what might be called the 'Nobel syndrome', a situation where scientists are called upon to give an opinion and pass judgement on any and every subject, however far removed from their field it might be.

The result has been that the Italian media as a group make no contribution towards the development of a 'culture of science' in the broadest sense of the term: the specialist journals administer their slice of the market with little imagination, leaving the coverage of the rest of science news to the sporadic and generally sensational attentions of the press and television. It is in this environment and with these considerations in mind that Hypothesis was founded, the first press agency in Europe dedicated exclusively to science, technology, the environment and science education. A group of communications professionals, for the most part journalists, took this initiative in order to go beyond the traditional concept of popularization, which inevitably results in the over-simplification of the complex material it deals with and is so often offensive to both those who practice it and those who benefit from it. The aim of Hypothesis is to provide clear, rigorous and, above all, interdisciplinary information: this is the only way to foster a true culture of science in Italy.

For more than three years now, Hypothesis has supplied articles, reports and features to a growing number of national and international media organizations, operating as a typical press agency. In parallel, it also functions as a publisher, producing books and multi-media presentations, as well as being involved in the organization of numerous conferences, seminars and round-table discussions. In most cases, the initiatives promoted by Hypothesis bring together all of these activities in one integrated effort.

The following two examples illustrate precisely what this project of integrated communication implies. Between 1990 and 1991, Hypothesis conceived and organized two international seminars: one investigated the relationship between mind and body ('Molecules and the Mind', at the Giorgio Cini Foundation in Venice); the other examined the nature of human and artificial intelligence ('Human, All Too Human', at the National Research Council in Rome), with the participation of many of the world's leading experts, including Karl R. Popper, John C. Eccles, Jean-Pierre Changeux, Roger W. Sperry, John R. Searle, Roger Schank, Michael S. Gazzaniga and Tomaso Poggio. The Hypothesis journalists, along with their colleagues from the leading national press, were involved in bringing the issues that emerged during the course of the seminar to the attention of the public (the final press coverage comprised more than 180 articles). With the collaboration of Giulio Giorello, philosopher of science, and Piergiorgio Strata, neurophysiologist, the papers presented at the seminar were collected in a volume intended for the general public, and published by one of Italy's leading publishers. All of this was in preparation for the concluding events of the Cortina-Ulisse European Award, the only Italian literary prize dedicated to science writing for the general public. The theme of the 1991–1992 competition, which was open to European authors and publishers, was the mind/body problem.

On the occasion of the Seventh International Conference on AIDS, which was held in Florence in June 1991, Hypothesis was selected to organize all the contacts with the world's press. The information and logistic support provided by the Media Centre went well beyond the usual press releases and press conferences: among the services planned and managed by Hypothesis were thematic guides to the more than 1000 papers read at the conference, a daily summary of the principal scientific results presented, seminars for journalists from countries in the developing world, a special information service for people with AIDS, technical and editorial support for television and radio crews, planning and assistance with interviews, and much more. The most important contribution, however, was the on-site writing and production of *Science vs Aids*, an English-language daily newspaper with 16,000 copies printed every night, whose articles were frequently cited by many publications including the *Washington Post* and the *New York Times*. The journalists attending the conference from around the world unanimously pronounced the media secretariat the best ever offered at an international AIDS conference. At the conclusion of the conference, the participating scientists in Florence were able to speak directly to the Italian public in a 96-page instant book prepared by Hypothesis in only four days; 220,000 copies were distributed with the weekly magazine *L'Europeo* as part of a national AIDS information campaign promoted by the Italian Ministry of Health.

The next step in the 'Hypothesis project' is the creation in Italy of a new source of science information, an experiment that may eventually be extended to the rest of Europe. We will soon begin publishing *Hypothesis*, a 32-page weekly tabloid, the first to be devoted to information and news from the scientific world, where science will not be relegated to a 'section' but will represent a point of view from which to interpret the changes that are transforming contemporary society. The launch of *Hypothesis* has been in preparation for the last year, and has drawn on the results of opinion polls and market research which revealed a demand from the general public for a publication of this kind. A survey conducted by Intermatrix provides a picture of the situation in Italy today: 43.1 per cent of readers with a relatively extensive educational background believed that today's press is too sensational in its science reporting, and 97.2 per cent said that they would like news sources to be more thoroughly investigated; 96.9 per cent thought there was a need for better qualified journalists. An interdisciplinary approach and more timely reporting were thought necessary by 85 and 87.3 per cent, respectively, of the survey respondents. All of these concerns are goals that *Hypothesis* has set itself in its coverage of the major issues that both more educated readers and the general public consider to be of vital importance: how to improve our health and the quality of life (cited by 45 per cent of respondents), what future lies in store for the Earth's environment (45.1 per cent), and the effect of scientific and technological progress on economic development (33.5 per cent).

Another key element of science information is the use of images, whose potential for communication is so often ignored. For this reason, Hypothesis is currently investigating the possibility of creating a data bank that will collect and organize the best of science iconography from Europe and around the world.

The current activities of Hypothesis are expected to take on an increasingly international dimension. It is hoped that similar initiatives will emerge in the rest of Europe so as to lay the foundations of a network for the exchange of high-quality, timely information on events and progress in the world of science. Hypothesis also aims to serve as a bridge between the EC and the Italian public, both through its own efforts and within the framework of the Community's Value II programme. In this way we hope to reduce another problem confronting the new Europe of science: the communication of information and policies adopted at the Community level which, at least in Italy, have trouble finding their way into the light.

Authors

Claudio Carlone, a science and economics journalist and communications expert, trained as a chemical engineer and biologist and began his career as a researcher in the field of advanced biotechnologies. He is currently director of Hypothesis, via G. Gioachino Belli 39, I-00193 Rome, Italy. Maria Vitale has worked for more than ten years in the field of science communication, as a journalist, editor and organizer of cultural events and exhibitions. She is currently head of cultural activities at Hypothesis.

Science without frontiers: a Media Resource Service for Europe

Jan Pieter Emans

The drive towards better public understanding of science needs to support efforts to build bridges between the world of science and the wider public. The media are an important channel in this respect: radio, television and newspapers constitute the main sources of information for most people.

Science and technology are integral parts of Europe's cultural heritage, in the same way as are art and literature. A basic level of public understanding of science will contribute to the public's ability to participate in debates on, for example, environmental issues, research in genetics or biotechnology.[1] Though studies have revealed that the public's knowledge and understanding of science may be limited, they also show that there is a real interest in scientific, medical and technological issues. Newspaper marketing surveys show that science scores higher than sport in reader interest, which might suggest that public concern focuses more on the degradation of the environment than on who won the Cup Final.

Living in a scientific and technological age has made public knowledge of these subjects increasingly important. Yet our knowledge or 'scientific literacy' is not always sufficient to make sense of complex matters. It is against this background that in 1985 the Ciba Foundation decided to launch the Media Resource Service. As a renowned international scientific organization for the promotion of medical and chemical research, the Foundation undertook to expand its aim of the improvement of communication within the scientific community to include a wider community – the general public. The Media Resource Service (MRS) offers journalists access to reliable background information from internationally recognized authorities – information which is essential if the public are to benefit from accurate and balanced science reporting.

The decision to try to improve the public's understanding by setting up a service for the media stems from the realization that most scientific information flows from the laboratory to the living room via the television screen, or reaches people as they read the newspaper on the way to work. Therefore it is important that the news and information media report scientific issues accurately. A story with a headline 'Breakthrough in Cancer Cure' may reveal in the text that such a development could in fact still be ten years away. Clearly access to reliable sources is a necessary ingredient of good science reporting. A study by Anders Hansen stresses that mutual trust lies at the heart of the relationship between journalist and scientist, and is of the utmost importance in the process of science reporting.[2]

These ingredients of accuracy, reliability and trust, together with excellence and expertise, form the basis on which the MRS operates. Over 5000 European scientists have already agreed to participate in the MRS venture. The techniques of recruitment used include literature searches in the scientific and mainstream press, advice from universities or other research establishments, contacts with scientific organizations, and increasingly secondary referral to scientists with greater expertise than those initially contacted from the MRS database.

Scientists who agree to join the register are asked to supply the MRS with a list of recent publications and a brief outline of their career history. They also give details of their areas of expertise, and this information allows identification by means of descriptors. In return the scientists are provided with a brief outline of how to conduct their contacts with the media. In this way the MRS tries to offer a service that allows journalists access to scientific excellence and to experts who can speak with authority and clarity when approached by the media.

The 'Service gives the names, contact addresses and telephone numbers of scientists on the database to those working within the media, from science writers to programme researchers. Help is provided for both news and feature stories or programmes. Sometimes a journalist simply wants background information. In other cases experts are required for interviews 'on the air'. A significant number of general news reporters or researchers who have no background in science contact the MRS when researching a story. Established science journalists, who usually have their own large networks of contacts, also phone the service when they find it difficult to locate a knowledgeable source.

Journalists have found their way to the MRS in increasing numbers over the eight years of its operation. All media enquiries are logged, and the pattern of calls is regularly analysed. The largest number of calls come from journalists employed in newspapers, magazines, television and radio. Freelance writers form another important user group. As a result of changes in the television industry, independent television production companies contact the MRS more often now than in the past. The local media are as much represented as the large national media. This is significant, since local media often lack specialist scientific staff and work to tight deadlines which do not leave adequate time for detailed research. One of the major users of the MRS is the BBC. Many BBC departments, from BBC News, the science radio and television programmes to the morning talk shows, call upon the MRS. A feature in the UK daily newspaper the *Independent* called 'The molecule of the month' is often written with the support of the service; and the *Economist* frequently seeks expert advice.

Most media attention focuses on subjects in the area of health and medicine which, during the winter months of 1992/3, accounted for 43 per cent of all calls to the MRS. The following examples give an impression of the wide range of requests the MRS receives: the production team of a British children's television programme wanted to know how to measure a human tongue; journalists from women's magazines and from the national UK press enquired about hormone replacement therapy after it was reported that some women had developed masculine secondary sexual characteristics; following the French government's move to ban smoking in all public places, questions were frequently asked on the medical dangers of passive smoking; and the medical, social and ethical definitions of brain death were the subject of many enquiries following a UK High Court decision to allow the life support system of a patient in a persistent vegetative state to be switched off. As the British winter set in, the MRS provided contacts for the many enquiries on seasonal affective disorder (SAD), flu and the common cold, and in the run-up to Christmas, we put the various media in touch with experts on snow, over-eating, office parties and reindeer.

The NASA operation to listen in space for extra-terrestrial life, which began in September 1992, prompted calls from journalists looking for experts to give their opinion on space technology, as did reports that the Swift–Tuttle comet could destroy the Earth in a collision in August 2116. The GATT talks in the United States generated much interest in rapeseed oil, particularly after experiments had shown it to be an effective alternative fuel source which was being used to power public transport. At the beginning of 1993 the oil spill off the Shetland Isles and the harmful environmental and medical effects of the oil received much attention from the UK national television and press. Here the MRS was able to provide a range of contacts covering the various aspects of media enquiries into the disaster – from marine safety and regulations to world-wide oil production and wastage figures.

Each year the MRS undertakes a survey to assess the contribution and effectiveness of its activities in the improvement and widening of science communication. The results show that a major proportion (over 85 per cent) of the experts suggested were new to the journalists – thus the MRS is increasing the number of sources available to the media. In many cases the journalist was also given the names of additional experts by the referred specialist. Around a third of the journalists questioned contacted the experts again directly after the initial contact through the MRS. Experts who saw or heard the item which resulted from their assistance were highly satisfied (90–95 per cent). This compares well with results from Hansen's study on science in the media, where the scientists interviewed were generally equally happy with their own contributions; they were, however, more critical about how their peers were represented in the media.[2]

The initial and continuing success of the MRS in the UK prompted the Ciba Foundation to expand the service to other European countries. The MRS database was therefore enlarged to include scientists in Austria, the Benelux countries, Germany, France, Scandinavia and Switzerland. Over the last two years an increasing number of media outlets in these countries have contacted the MRS. Among regular users are *Science et Vie*, *France 2*, *Die Zeit* and *NRC Handelsblad*. These requests represent something which might be called 'cross-border science writing'. But the European expansion of the service took off more slowly than initially expected. The underlying factors for this might stem from what Pierre Fayard describes as 'the domination of Anglo-Saxon science information sources', and the lack of sufficiently well-known broader European sources of scientific and technological information.[3] However, once journalists find their way to the Media Resource Service they invariably become regular users.

The Ciba Foundation also lends its support to various other endeavours in the world of science communication. A fellowship administered by COPUS, the Committee for the Public Understanding of Science, allows scientists to spend up to six weeks in a media environment in the UK. Its aim is to give first-hand experience of the conditions and constraints which face working journalists, and to better equip the scientists to communicate their work to the general public, their students and their colleagues. More recently the German Max Planck Institutes started two-week placements for journalists in their Munich laboratories, to give them better insight into the process of scientific research and method. Many European universities now offer courses which teach science communication either as part of a degree or as a full degree course, on the basis that academic training in a particular subject on its own does not necessarily mean that scientific knowledge will be accurately communicated by those who graduate.

The latest shoot on the blossoming MRS tree is a twice-yearly newsletter which informs scientists and journalists about the activities of the Service and also aims to report on the varied efforts elsewhere in Europe to improve the communication of science to a wider public. In the first issue the Director of the Ciba Foundation, Derek Chadwick, quoted Thomas Jefferson who wrote 200 years ago: 'I know no safe depository of the ultimate powers of the society but the people themselves, and if we think them not enlightened enough to exercise that control with a wholesome discretion, the remedy is not to take it from them, but to inform their discretion'. That is exactly what the MRS has been trying to do for science. Chadwick added that its function is not to usurp the journalist's role, but rather to act as an 'honest broker', forging contacts between journalists who want accurate information and those scientists able and willing to provide it. Then, at least, readers, viewers and listeners will have first-rate scientific information on which to base their judgement of the scientific enterprise and its effects.

References

1   Durant, J., 1991, Why scientific literacy matters. *Science Communication in Europe* (London: Ciba Foundation), pp.70–73.

2   Hansen, A., 1992, What if there are multiple intentions? Journalistic practices and science coverage in the British press. Paper presented to the Annual Meeting of the AAAS, Chicago, 6–11 February.

3   Fayard, P., 1992, The development of science reporting in the European daily press. *Science and the Media – A European Comparison*, edited by K. Zerges and W. Becker (Berlin: sigma) pp.97–104.

Author

Jan Pieter Emans is Information Officer at the Ciba Foundation, 41 Portland Place, London W1N 4BN, UK.

Daily science in the European quality press

Pierre Fayard

When studying the evolution of the coverage of scientific and technical information in the major national press in Europe, three major conclusions stand out. We notice firstly a widespread explosion in the coverage of this type of information in the Community countries studied; secondly an increasing specialization by scientific journalists; and finally the inability of Europe to control the network for making known and circulating scientific information.[1] Using current events which are on the whole international, scientific journalists writing in the daily press continue to work 'to order', and in this way the entire cultural diversity in Europe expresses itself, and is now developing.

First negative aspect: Eastern Europe is moving in the opposite direction

However, the East is moving in the opposite direction. The opinions gathered concerning the Czech daily newspapers *Malda Fronta Dnes* and *Lidové Noviny* reveal by contrast what the norm is in the West. In Prague each period of political liberalization (1968, and then the Velvet Revolution) is accompanied by a reduction in the amount of scientific and technical information in the press. During periods of hard-line communism, the coverage of the exact sciences (the sciences based on mathematical reasoning) represented for journalists an area of freedom since half the subjects dealt with remained Soviet. But since the ideological control eased off, journalists who until then had been restricted to the sciences were suddenly investing their energies in the areas of philosophy, history, culture and even politics. In the East, today, not only is the shortage of international scientific information considerable, but there is also very little awareness of the importance of science. With the exception of personal contacts with members of the Academy, the material which journalists have at their disposal comes almost exclusively from the important North American daily newspapers and occasionally from the magazine *Nature*. West European countries ignore their East European neighbours, and so the sources of information from the East have a tendency to dry up.

Western Europe: coverage explodes as public curiosity grows

In the West the coverage of the sciences increases in parallel with the growing interest of the public (see the table). The increase in the amount of coverage of the sciences established itself in the 1980s and the trend continues today with the development of environmental themes dealt with by scientific journalists. In Great Britain the 'quality press' devotes the least space to scientific and technical information. In *The Times* this has reduced drastically in parallel with a fall in receipts from advertizing. For the same

The science supplements in the dailies studied.

Belgium	*Le Soir* – Brussels, broadsheet, 160,000 copies. Title: 'Sciences et Technologies', 1 to 2 pages weekly, 1 journalist.
	de Standaart – Brussels, broadsheet, 80,000 copies. Title: 'Kultur & Wetenschaft', to 1 page daily, 3 journalists.
Czech Republic	*Mlada Fronta Dnes* – Prague, tabloid, 400,000 copies. No supplement, 1 journalist.
	Lidové Noviny – Prague, tabloid, 150,000 copies. No supplement, 1 journalist.
France	*Le Figaro* – Paris, broadsheet, 400,000 copies. Title: 'La vie scientifique et médicale', 1 page daily, 8 journalists (including medicine).
	Le Monde – Paris, tabloid, 500,000 copies. Title: 'Sciences et médecine', 3 pages weekly, 4 journalists.
	Libération – Paris, tabloid, 250,000 copies. Title: 'Eurêka', 6 to 8 pages weekly (included supplement), 5 journalists.
Germany	*Frankfürter Allgemeine Zeitung* – Frankfurt, broadsheet, 400,000 copies. Title: 'Natür & Wissenschaft' – 2 to 4 pages weekly, 6 journalists.
	Die Welt – Hamburg, broadsheet, 230,000 copies. Title: 'Umwelt & Wissenschaft' – 1 page daily and 1 weekly, 4 journalists.
	Süddeutsche Zeitung – Munich, broadsheet, 550,000 copies. Title: 'Umwelt & Wissenschaft' – 3 to 4 pages weekly, 3 journalists.
Great Britain	*The Times* – London, broadsheet, 440,000 copies. Title: 'Science', ½ to 1 page weekly, 4 journalists.
	The *Guardian* – London, broadsheet, 400,000 copies. Title: 'Science', 1 page, weekly, 1 journalist.
	The *Independent* – London, broadsheet, 390,000 copies. Title: 'Science & Technology', 1½ pages weekly, 2 journalists.
Italy	*Il Corriere della Sera* – Milan, broadsheet, 800,000 copies. Title: 'Scienze', 4 pages weekly, 3 journalists.
	La Stampa – Turin, broadsheet, 450,000 copies. Title: 'Tuttoscienze', 4 pages weekly, 3 journalists.
	La Repubblica – Rome, tabloid, 1 million copies. Title: 'Il Mercurio', 1 to 2 pages daily, 4 journalists.
The Netherlands	*NRC Handelsblad* – Rotterdam, broadsheet, 250,000 copies. Title: 'Wetenschap & Onderwags', 6 pages, 4 journalists.
	de Volkskrant – Amsterdam, broadsheet, 340,000 copies. Title: 'Wetenschap', 4 pages, 7 journalists.
Portugal	*O'Publico* – Lisbon, tabloid, 75,000 copies. Title: 'Cienca e tecnologia', 1 page daily, 4 journalists.
Spain	*El Pais* – Madrid, tabloid, 400,000 copies and 1 million on Sundays. Title: 'Futuro', 8 to 12 pages, 2 journalists.
	La Vanguardia – Barcelona, tabloid, 200,000 copies. Title: 'Cienca y Technologia', separate supplement of 16 colour pages, 8 journalists (including medicine).

---

reason, the sixteen-page science supplement of the Portuguese *O'Publico*, even though it has been chosen by readers as the most popular publication, has had to be reduced in size due to the lack of advertizing.

At the other extreme, the most prolific country is Spain where in Barcelona *La Vanguardia* increases its sales by 10 per cent on the day when its colour supplement comes out. In Milan the circulation of *Il Corriere della Sera* increases by more than 30,000 on the day its science supplement is published. Every six months *La Stampa* publishes 25,000 copies of a selection of pages from 'Tuttoscienze'. In *Le Monde*, two studies carried out with a five-year interval emphasize that 'Sciences et Médecine' still has a readership approaching 30 per cent, which nearly equals that of the daily economic pages. 'Wetenschap & Onderwags' in the Rotterdam *NRC Handelsblad* is the most popular of the newspaper's supplements.

In Germany, the science sections in the three newspapers studied all have 'nature' or 'the environment' in their titles in association with 'science' or 'research'. In the Netherlands, education is linked with science in both the titles and the substance of the supplements themselves. In Italy, the great 'Roman and national' institution *La Repubblica*, which is a national newspaper but not dependent on the Vatican, prefers not to separate science and culture, to avoid the marginalization of this kind of information. The same editorial line is followed in the offices of the Flemish *de Standaart*. The French and German daily newspapers *Le Figaro* and *Die Welt* both reserve space for science every day.

The existence of weekly supplements does not exclude the daily coverage of scientific information linked, according to the journalists, to 'current issues read about on public transport and which people want to talk about over a coffee. Articles in the supplement are the subject of evening reading where an increased popularization could take place'. The systematization of the 'quality' science pages in the 1980s has granted journalists a certain freedom. This recent phenomenon is the result not only of the readers' expectations, but also of a specific way of dealing with science. The key ideas in journalism are current affairs, the readers, loyalty and clarity of content. It is not a coincidence that today this type of approach, applied in certain areas of the Cité des Sciences et de l'Industrie de La Villette, is the subject of great interest on the part of a few of the prestigious European science museums.

Reporters specialize and take outside collaborators

The age of 'general' scientific journalists, who were capable of covering all the disciplines, seems to be over. Jacques Poncin of *Le Soir* evokes the image of his predecessor as 'an old and very respectable gentleman, Flemish, but fluent in French and who dealt with all the sciences in one regular column'. With increasing numbers of staff, journalists specialize in specific disciplines. When the know-how is lacking, freelance journalists are called in, and in certain daily newspapers freelancers write up to half, or even more, of the supplement. Occasionally it happens that the scientific community itself will be called in, and in this case it is the choice of the editor. But the journalistic orthodoxy which is widely professed excludes, through a sense of what is realistic, calling in researchers to write articles, except in the cases where they succeed in writing 'journalistically'. Not one scientist puts their name to articles in British supplements, in France and Germany it is very rare, and in Portugal it happens occasionally. It is on the other hand frequent in Spain, Italy and the Netherlands.

Among the common trends, the high level of training acquired by scientific journalists is clear. Half of those responsible for the eighteen West European daily newspapers possess a doctorate in science, as do other permanent employees. Malen Ruiz de Elvira at *El Pais* is the only one to combine both engineering and journalistic studies. The other journalists often have a masters degree in exact sciences and less frequently in human and social sciences. It is indisputably in the field of scientific journalism that the nature of the initial training acquired corresponds best to the current issues dealt with, which is more rarely the case in the fields of sport, politics, music, or even economics.

Following the examples of other fields of journalism, there is, however, still room in the field of scientific journalism for the self-taught and for those who have changed fields. One of the permanent members of staff at *The Times* was formerly a pop singer, but then 'nobody's perfect'. Italy stands out due to the dominance of people transferring from the humanities among its teams of journalists, the exception being Giovani Caprara of *Il Corriere della Sera*, an aeronautical engineer. 'In Italy, the most important thing is to be able to master the language, to have access to news and have reliable sources' says Marina Verna of *La Stampa*, a former Latin teacher who is fluent in French, German and English, and specializes in musicology. In the sciences, 'machismo' is more of a Northern European phenomenon: Spain, Italy, France and Czechoslovakia are the only countries where those responsible for the supplements are women.

Linguistic barriers and the culturally mute

Like researchers, all scientific journalists read and speak English. The most polyglot are still those who speak Romance languages including French, and some are even fluent in a fourth language as is the case for the Portuguese and the Italians. At *O'Publico*, knowledge of English and French is a prerequisite for becoming part of the sciences team. In Northern Europe, if we put aside the 'comment allez-vous now let's speak English if you don't mind', or the 'buenos dias I've been to Spain on holiday', the omnipotence of English stands out. Because of their size, the Netherlands and Portugal are both very open towards other countries. For the British, the natural outside link is with North America, which alters their geographical perceptions: for them the English Channel becomes a veritable ocean and the Atlantic barely a river. But Nigel Williams of the *Guardian* recommends, as often as possible, the reading of the Munich *Süddeutsche*'s science supplement.

In terms of sources of information from other European countries, in Spain, Italy, Portugal and Belgium journalists mention publications from the CNRS (the French National Centre for Scientific Research), sometimes from the Pasteur Institute, from the INSERM (the French National Institute of Health and Medical Research) and the embassy services. In London, the dossiers from their CNRS offices are mentioned, as is a similar service from Germany. In the Netherlands, official German sources, in English, are also used. In addition to the omnipresence and necessity of English, to which French is added in the South and, most probably, German in the centre of

Europe, the other cultures and their languages are condemned to silence on an international level, although Spanish appears to be gaining importance.

Scientific results: can European speak unto European?

The almost exclusive use of the Anglo-Saxon language means that the primary scientific press, the essential source of information for scientific journalists, is concentrated across the English Channel and across the Atlantic. This Anglo-American supremacy influences the very definition of current scientific issues, and of the 'agenda setting'. By means of their 'preprints' the primary weekly journals have at their disposal an essential basis for directing the agenda of current issues. The daily publications which subscribe to these journals can use the information for articles which they can publish on the same day of issue as the original sources, which amounts to giving great weight to *the* piece of news of the week: the one which the journal has selected and chosen to sell.

This mechanism has the objective support of the major North American dailies like the much-mentioned *New York Times*, the *Washington Post*, their joint publication the *International Herald Tribune*, as well as the *Los Angeles Times*. Journalists obtain far higher quality information from these publications than from the large international press agencies like Reuter, which is top of the list, followed by Associated Press, the Agence France-Presse for French speakers, United Press and finally Tass, which is mentioned only in *Le Monde*. Journalists who deal with science and technology in the major North American dailies are real specialists, which means that their articles are accurate, well-researched and reliable, and in terms of quality cannot be compared to the press releases of their non-specialist colleagues at the large agencies.

Thanks to its mere logistic effect, this mechanism of selection and circulation of scientific information, whereby the choices of the primary journals, the major North American dailies, a few international agencies and the work done in secret by press offices at the major US universities all converge, functions like a 'network' in the literal sense of the word. It is a thread – a web – which is at work. Thanks to the geographic localization of the centres of decision-making within this web, access to the 'front page' is easier if you work in the English-speaking world. The result of this is that for results of research done in Germany or Italy to become known in France or in Spain, they have, almost compulsorily, to pass through the USA or Great Britain.

European scientific journalists have great difficulty in freeing themselves of this logistic pressure, because each editorial staff is in competition with the other dailies in its country. Not knowing *the* piece of news of the week could make one newspaper look incompetent if the other national dailies are talking about it. What is odd is that those most hostile towards this Anglo-Saxon hegemony are not the French as we might have expected, but the Germans and the Italians, who denounce, in very blunt terms, the '*mafia* which gives priority to sales rather than to scientific content, and establishes national preferences'. Though science is by nature international the monoculture of its exploitation is establishing inequalities and is favouring only one way of doing research and of presenting the results. Europe seems unable to control either the network to make known the research carried out, or the flow of news.

Towards European scientific and technological information

For a long time North American themes and editorials had more or less imposed the example that Europe was to follow. In several countries a number of articles were regularly translated and published. Now the trend is shifting and the US articles are regarded as 'too infantile' in the Netherlands or 'badly constructed' in Italy. But there seem to be few drawbacks should the Americans or others want to translate articles from the Dutch or Italian ... . The present maturity of scientific journalism in Europe comes partly from its rapprochement with local scientific communities which provide the means of breaking free from the American grip. A tendency of the primary magazines to 'nationalize' the current important issues establishes itself by means of an increased cooperation with researchers. Dailies which are in direct contact with geographical entities such as Catalonia, Piedmont or the Netherlands collaborate intensely with scientists who write their own articles. This is a new phenomenon.

The more liberal dailies like *The Times*, *Le Figaro* and *Die Welt* do not challenge the fundamental principles of the mechanism of selection and circulation of scientific information. Rather, they see it as an excellent means of obtaining the best research news. In addition to this, they proclaim to be more interested in what is happening in the USA and Japan than in what is happening in Europe. The more Community-oriented dailies like the *Independent* or the *Frankfürter* regret that there is not a similar system which would allow European journalists to be informed of what is happening in neighbouring countries, and to know where the centres of excellence are in terms of science. It is still easier today to find out what is happening in American universities than it is to find out about those in Europe.

The official information from Brussels is unanimously classed as useless as it is 'too political and it focuses on projects and programmes whereas what we are looking for is results'. The communication from universities and European research centres is regarded as nonexistent or clumsy, the exception being a few British universities. Trips from country to country typically constitute a practice above all on an individual level, based on a list of addresses, as there is practically no organized flow of communication. Government services in France have a reputation for being efficient and are well known in England and Spain, and those in Germany are held in similar esteem in England and the Netherlands. The trips organized by the EUSJA (the European Union of Associations of Scientific Journalists) are still not enough to reverse the trend.

Die-hard diversity

In addition to this problem of the actual physical circulation of information, there is also the problem of cultural open-mindedness of journalists who judge information and the methods of obtaining it according to their national criteria. Rainer Flöhl from the *Frankfürter* does not understand why the French give greater importance to lunching with the bosses than to providing useful information. He now refuses invitations from France after one such experience. When referring to the EDF, the French electricity company, the Dutch journalist Robert Biersma says that it is '100 per cent French-speaking, but when it concerns research into space, then it's a different matter'. English journalists profess to be dumbfounded by the lengthy editorials in Germany and the Netherlands. In Germany, they do not understand the absence of contextualization in

English articles. The Dutch think the Germans are serious but boring. The Belgians lament the fact the Spanish do not answer their mail. As far as the Portuguese are concerned, they bitterly note that they are not respected by European laboratories. If we add to this the stereotypes and the weight of past historical events we can see that the European traditions of a mutual lack of understanding represent an enormous stumbling block which can be mobilized with unexpected ease – unlike the self-questioning necessary to make the effort to understand others.

As an honourable native of the land of Descartes who considers the law of the 'excluded middle' as a law of nature, I was extremely surprised when a British journalist expressed two contradictory opinions on the same question only half an hour apart and without the slightest embarrassment. Not in the least confused by our surprise, her Gracious Majesty's subject admitted that his original response was right although he would allow himself to have a contradictory attitude in the future. So would 'To be *and* not to be' become a trait of British culture, not strictly limited to European membership, in this country which feels both 'inside *and* outside' Europe? Among the widely accepted paradoxes Italy is exemplary, since there journalism is considered to be a liberal profession, but to obtain the status of journalist one has to be a salaried employee of a magazine or newspaper.

Journalism: working with the difference

On principle journalists want to grasp everything which allows the establishment of direct relations with researchers, and they want to acquire information prior to the publication of research results. The ideal solution for them would be to develop a capacity to anticipate the contents pages of the primary journals via knowledge of and regular contact with the eminent centres of scientific research in Europe. It is also why they call for increased professionalism in terms of what the Universities communicate to them. They admit to being more curious about themes and the way they are treated by their colleagues from other cultures than they are about translations.

Journalism is a result of diversity and in this regard, Europe represents an ideal location. 'Actually it is the cultural aspect which is interesting,' maintains Rainer Flöhl. 'Certain ailments are illnesses in some countries and not in others. If everything was the same, there would be no reason to write at all – it is the diversity which makes it interesting'. Franco Prattico of *La Repubblica* thinks that Europe has a specific way of looking at fundamental research. Nigel Hawkes of *The Times* 'would like to know the German laws on genetic engineering'. Dominique Leglu of *Libération* suggests 'the grilling by several journalists of key scientific figures about current issues'.

Time seems to be working in favour of a better circulation of scientific information in Europe. Tom Wilkie of the *Independent* clearly identifies the harbingers of a sort of 'Renaissance', a geopolitical revolution in the sense that 'Alcatel is bigger than ITT, Bell Communication is no longer the leader, and the Germany's economic influence is considerable'. In answer to the question of who could make the most effective contribution to the coverage of scientific activity across Europe, he replies by presenting a half-inch thick A4 document. Therein, succinctly presented, were the addresses and others details of the eminent centres of scientific research in Great Britain. Tom Wilkie says that he is in favour of such a work tool on a European level which would provide him with a 'mental map, broad picture of what is worth [*sic*] in Europe and in each

European country.' It is the journalists' business to find out the rest.

Though it is still easy to profess one's faith in Europe, deeply rooted in its cultural environment and convictions, it is quite another matter when one is faced with other ways of communicating and perceiving others, with other value systems, and other ways not only of working but also of laughing, eating and drinking ... *and* with accepting them, while remaining loyal to what one appreciates and what one believes in in one's own culture. The construction of Europe can only be achieved by means of more complexity, of knowledge and tolerance, of intelligence and open-mindedness, and not by means of a process of simplification which can barely be applied in terms of economic norms. The existence of a single language in Europe, like a single culture, could well transform the Community into a potential Yugoslavia, since the English will always drink lukewarm beer and the Italians strong coffee, Germany will always be well organized, life will for the most part continue to be nocturnal in Spain and the conviviality of business lunches and dinners will remain, thankfully, an essential ritual in France.

Reference

1   This article summarizes the conclusions of a study which was carried out with the help of the European Commission, the French Ministries of Research and Technology and of Culture and Communication, the Cités des Sciences et de l'Industrie, Paris, the University of Poitiers and the PCST network. The study was published in Fayard, P., 1993, *Sciences aux quotidiens* (Nice: Z'éditions).

Author

Pierre Fayard, a former science journalist, is head of the LABCIS (Research Laboratory on Science Communication and Strategic Information) at the University of Poitiers, 40 Av. du Recteur Pineau, 86022 Poitiers Cedex France. He was the initiator and first coordinator of the PCST Network (Public Communication on Science and Technology) and has written many articles and several books. A full account of the study presented in this paper is available in *Sciences aux quotidiens, l'information scientifique dans la presse quotidienne européene* (Nice: Z'Edition, 1993).

Eurometrics: should we have taken the measure of Europe?

Bruno Latour and Mickès Coutouzis

In 1988 we had the rather crazy idea of celebrating the new, united Europe – and in the most difficult manner: by starting from the idea of technical standardization. We wanted to hold a giant exposition which would begin in all twelve countries at the same time on 1 January 1993, and the electronic connection of the 12 sites would illustrate the difficulty, the importance and the beauty of the building of Europe. It would be, in short, a colossal exercise in the relations between science, culture and politics. We were somewhat naive and our project, caught up in the agony of the collapse of the Berlin Wall, dissipated – like Europe? – after an enormous design effort that had lasted three years and had united eight European teams. Only Greece, the cradle of thought and civilization, is continuing with a 'Eurometrics: Greece' project, which opened in 1993. This article recalls the intentions of the project and offers a prefiguration of the exposition. Only those who have a narrow vision of Europe will view this project and this exposition as a megalomaniac dream. At the time, we thought that such an accomplishment deserved more than some self-conscious speeches, a symphony concert and a release of balloons.

The philosophy of the project – in 1990

The completion of the Open Market in 1992 requires an effort of harmonization by all the countries of the community – an effort unequalled in history for its scope and speed. This effort continues, and expands the technical work by which European nations and states have unified their own territory. Europeans cannot enter the Open Market collectively without celebrating and accentuating this event by delving into the diversity of national technological histories.

The work of normalization, standardization and harmonization is not accessible only to specialists and technicians. It must, and it can be, understood by the great majority of people. A whole technical culture must develop which will communicate the importance, the cost and the relevance of the 'measure' of Europe. It is a question of communicating, of making equivalent or uniform, national networks the different controls of which have, up until now, constituted the specificity of the European states: transportation networks, telephone or information networks, metrological channels, regulation sizing, networks of test laboratories, administrative formalities, statistics and accounting systems, followed by the transporting of people and goods ... . It is a matter of doing for all of Europe what each country has done throughout its long history to unify or harmonize its own national territory at the regional and provincial levels. This specialized technical work, which is often unrewarded but of immense scope, remains little known to the general public, or is only grasped in snatches, depending on the news media and the interests of specialized sectors.

This situation poses a problem for the completion of the Open Market of 1992. It has taken decades, even centuries, to harmonize or unify the national territories. The collective agreement on the metre, for example, was not signed until 1875, and then only by a few countries, almost a century after its definition by the members of the French Convention. The consumers and citizens of the future unified Europe cannot be set before a completed Europe. Europe cannot build itself unless the larger public understands the difficulties and the techniques of the harmonization.

Why? Because a specific problem presents itself to Europe: it is impossible and undesirable to render technically uniform an old continent made up of 320 million inhabitants with nine different official languages, whose regions and provinces are more and more significant, and whose history is marked by differentiations and cultures. No one is proposing, for example, the creation of a common European language, any more than anyone would consider it possible to create a single laboratory for testing and certifying all European products. For the first time in history, we must think in terms of a technical unification without uniformity and without the smoothing over of local specificities. The goal is to harmonize, to make coherent, and to make equivalent the new Europe of the Twelve, but without losing the specificities, the identities, the differences. This struggle between uniformity and harmony is not cultural but technical, and it is everywhere: it is found both at the top, among the specialists of normalization with the 'new approach' of mutual recognition of testing and certification, and at the base, among the consumers, in the decisions they make in their choice of products. What good would it do to unify the norms if consumers did not accept the seals of quality of the other countries? What good would it do to legally permit the free movement of goods and people if nationalism made this movement impossible in practice?

The two channels that have favoured technical unification over the course of history – that is, national centralization on the one hand, and market forces on the other – will not suffice to make a citizens' Europe. Whether it is a question of norms for the transfer of information in computer networks, of security regulations in power stations, or of statistical categories permitting the calculation of the rate of unemployment, it is necessary in each case to take into account national or regional specificities and, at the same time, to harmonize. This task of delicate selection must be understood by the majority because it is an essential ingredient of culture.

The extraordinary mobilization of the Open Market in 1992 and the giant task of the normalizers and standardizers must be paralleled by an elevation of the technical culture. Instead of proposing to Europeans the not always exhilarating role of being interchangeable consumers in a uniform market, we must all be the ones who choose, collectively, the degree of technical harmonization of the new whole to be created: what our ancestors did for the nations, we are ready to do again for Europe, on the condition that we be given the means to understand the methods and the stakes. If the culture of Europeans is to be common it must also be technical.

The technical harmonization of Europe is recognized by all, including those who do it every day, as a job that is at once strategic and very difficult to explain.

Up to this point, the solution has consisted of sensitization campaigns for manufacturers and decision-makers, and resolution of the problems by sectors. The public is left out; they see the results (almost always esoteric), but they do not see the stakes involved – often enormous, the negotiation process (always exciting), and especially the financial, intellectual and organizational cost of the collective acceptance of harmonized measures. The European public has never had the opportunity to size up in its entirety, with all of the risks and in historical context, the constructing of harmonized technical networks – of a new social body, in fact, that makes Europe into an irreversible reality.

In order to make the task of technical harmonization interesting for the general public – a task that may seem forbidding at first – there must be a new philosophy of measures: instead of starting with the equivalence between sizes, we must start with the task of making equivalent; instead of imposing the results of the European decisions, we must sensitize the public to the often conflict-ridden process of harmonization; instead of assuming that the technical data, the evidence, tests, norms and tolerance levels are naturally universal and unambiguous, we must show that these are in fact often fragile constructions, which are always costly and which are the result of difficult negotiations and arbitrations. The construction of Europe and the construction of normalized data and equivalencies go hand in hand. The general public must understand both the advantages and the inconveniences of the two constructions.

In order for the general public to perceive this work of harmonization, and for European technical culture to profit from it, the vision of the Open Market must not be atomized, but must be treated as a single unity covering everything from metrology to quality, from normalization to the free movement of goods and people.

How can we establish a universal time? How can we harmonize the national finance systems? How can we make quality control testing equivalent in all countries? How can we make university diplomas compatible? How can we decide on common safety regulations for products? How can we have a European currency? All of these questions are common, whether they revolve around legal metrology or the more general process of standardization, whether they concern the most scientific subjects or more economic and administrative matters. And all turn on the construction of chains of equivalence, on the maintenance of costly networks, on agreements, rules, instruments, devices, and institutions. It is the total of all of these things that must be explained in a coherent manner.

There exists in Europe, among the historians, the economists, the sociologists, the philosophers, the pedagogues and the metrologists, an immense reservoir of investigations, studies and arguments on the history, economy and philosophy of this enormous task of modernization, rationalization and scientifization of territories, states and peoples. It is not an exaggeration to say that since the eighteenth century, all of the great thinkers and historians have studied, in one form or another, this enormous scientific and technical transformation that makes us both modern and European: from Montesquieu to Jurgen Habermas, through Adam Smith, Emmanuel Kant, Max Weber and Michel Foucault. It is this task that we must continue, build up, enlarge and make perceptible to the general public engaged in the construction of a new identity. Only then will we be able, thanks to the work of technical harmonization, to answer the big question: what does it mean to be European in 1993?

To the mobilization of the institutes of normalization and standardization, and of manufacturers, economists and states, must be added the mobilization of those intellectuals, academics, pedagogues and researchers who work at understanding the multiple ties between the European sciences, technologies, economies and societies of yesterday, today and tomorrow. There must be an immense investment of research made in the human sciences and in pedagogy to create, along with the European Technological Area, a Technological Culture that will accompany, permit, democratize and justify this area. We – academics, intellectuals, manufacturers, researchers, pedagogues, statesmen and various leaders in the completion of the Open Market and in the building of Europe – have decided to launch this effort under the title of 'EUROMETRICS: Taking the Measure of Europe'. This effort will express itself, firstly, through the continued mobilization of the sectors involved in research on the multiple links between sciences, technologies, economies and societies. Secondly, a European event will take place on the enforcement date of the Open Market, which will create, simultaneously in the 12 capital cities, a network of expositions and shows.

What is the essential pedagogical message of the exposition? To show the work of those who are carrying out, and who have carried out the technical harmonization; to make clear the cost of this harmonization and the cost of its absence; to make the general public more aware of the problem of measurement.

Prefiguration of an exposition – as imagined in 1990

It is 1993. Here we are in front of the great hall of the exposition: 'EUROMETRICS: Taking the Measure of Europe'. The same exposition is taking place at the same time in the eleven other countries, but with a number of interesting differences: some are taking place in large, national, technical museums, while others provided the occasion for the creation of a museum of sciences, technologies and industry. All provided an opportunity to dust off and to show off precious national collections on the history of metrology. All were also the result of an intense historical research effort on the national history in its relations to measurement. Unlike many expositions of this kind, however, each demonstration and each game demands time and the investment of prepared teams. Hence the special atmosphere, which is both that of a European game like *jeu sans frontières* made up of teams and challenges, and that of a large museum of technology.

The first thing that strikes the visitor to Eurometrics is not the crowd pressing into the exposition, but rather the instruments by which the crowd finds itself being reproduced and analysed. Large screens show simultaneously the images of the crowds participating in the eleven other expositions. In addition, the crowds are constantly studied by devices that measure their height, their weight, and which display, in real time, using immense histograms, the size and the movement of the crowds not only at that exposition but also in the twelve countries simultaneously. For example, it is possible at any given moment for a Greek to see the average height of Danes or for a Portuguese to find out about the Luxembourg public's perception of technological dangers. The visitors are bombarded by children carrying out investigations for various school projects (on the perception of dangers, on diplomas, on the fondness for pets, etc.) and who will later, through the intermediary of large video screens, compare their results and measurements with those of other groups of children in other countries.

But even more striking is the fact that the atmosphere of the exposition is neither cold nor glossy despite all the machinery. On the contrary, all of the possible measurements taken are clearly displayed. All of the measurements for the signs, the partitions, the clocks, the recording units, the video control rooms and all of the amounts and totals are visible. Even the price and the bar code on each unit has been preserved. Instead of being erased and hidden, the measurements are themselves the object of the exposition. The metrology stations have their own laboratory at the very centre of the exposition and must, in the difficult atmosphere of the crowd and the heat, maintain the precision necessary to coordinate the immense network of the twelve expos.

A large part of the expo turns on metrological culture. How is a measurement taken? How is a tolerance level calculated? How can an experiment be replicated? How can precision be achieved, and at what cost? At one stand, for example, anyone able to guess his or her own weight, to the milligram, is promised their weight in gold. There is a queue of people ready to take up the challenge, and they soon realize how difficult it is to limit the margins of error. At another stand, visitors at the other expositions are asked to match exactly, on the screen, the weight of the visitors at this end. They must choose from among the crowd those who will make an exact match, but since the people at the other expo at the other end of Europe are doing the same thing, the task becomes complicated and the equilibration of the great balance beam (electronic, of course) proves difficult.

Many of the stands make extensive use of the simultaneous linking of the twelve expos to play on the relativism of Europe – that is, to build awareness of the specificities that must disappear and those it would be advisable to keep. 'Truth is relative.' Nurses of one country compare their diplomas and their schedules with nurses of another country. The game of ethnomedicine permits a comparison of what the different countries believe themselves to be the most afflicted with, and of how they care for themselves. However, as always at the expo, the games also compare the methods of calculation in the different countries. For example, when there is a figure for unemployment in a given country, or for the number of cases of lung cancer, or for the risk of radiation exposure, these figures appear a second time calculated according to the method of another country. The disparities are always compared. A televised game allows a visitor at one expo to ask a dentist or a mechanic in another country if he or she would like to immigrate. The enthusiastic or embarrassed responses of the person interviewed outline the felicitous and unfortunate specificities of Europe. Another game, for children, is a 'career' game where the different professions must be mimed by overcoming the barrier of gestures, so different from one culture to another. Of course, the exposition deals with anything concerning the stereotypes that Europeans have of each other: the portrait game displays each day the pictures of thirty inhabitants from different countries and asks the visitors to decide which one most resembles the Danish 'type', the Greek 'type' or the French 'type'. The idea of 'types' is thus invalidated and visitors learn to be suspicious of simplifications.

Obviously, money is not used at the expo. In its place, amounts are converted into the national currencies that are used for each stand – since the other eleven countries are represented. But of course the value of the money changes throughout the day in relation, for example, to the number of visitors, and this variation affects all of the electronic games at all twelve expositions using the currency. This variation certainly complicates the games, especially the very popular electronic monopoly for twelve, where the goal is still to buy hotels and houses but on streets that each belong to a different European capital (with its specific rental legislation, its money, its language and its rules of play ...).

One of the most spectacular aspects of Eurometrics is the presence at the expo itself of testing machines from laboratories. The machines pull, pierce or ink paper, pound on chairs, smash cars, bang into cement partitions, smoke cigarettes, etc. But these machines are not only amazing, they are also the object of a challenge. There are prizes for visitors who are able to put the testing centres of different expos into competition on the same product. Thus, beside each machine can be seen the video rerun of all the other machines of the Twelve with all of their differences and similarities. The challenges are especially numerous in the farm produce section, where the visitors never tire of comparing the results from different laboratories on the percentage of fat in cheeses, or the percentage of alcohol in wines. At each site, the experiments illustrate the difficulty of measuring, and the reasons for the observed disparities. Little by little, the importance of the tests and certifications to the problem of sampling is understood, as well as the usefulness of controls. Bit by bit the visitors' respect grows for the technical difficulties. The most popular game from this point of view is the game of surveillance against major technological disasters. All of the large European institutes have an annexe of their laboratories right in the expo and, by simulating ecological alerts, they compare the variety or the similarity of the reactions and the data, and always insist on the difficulty of measurements.

But what most surprises the visitors is the on-going integration into Eurometrics of the history of their own country and that of Europe. In the great hall of the provinces, for example, an immense working model shows the movement of goods and people at different periods in the national history. The spectator is astounded by the slowness, the number of interruptions, the ruptures, and the burden of formalities that had to be dealt with, fifty, a hundred or two hundred years ago, in order to pass from one province to another (what the modern visitor now considers to be a single area). Another working model represents Europe before the abolition of customs tariffs. Elsewhere, a series of objects and of games reminds the visitor of the diversity of weights and measures and the intensity of local cultures. Children make a chain and try to pass along a certain quantity of grain, but each child is using a container corresponding to those used in a medieval city. The difficulties of transport appear clearly. Another hall, called The Babel of Regional Languages, shows the dynamism of regions that is favoured by the Europe of '92. The information stands, for example, are often in a language other than the official language. As always at the expo, the organizers use all of the visitors' own reactions to help them see the difficulty and the importance of the standards. Telecommunications certainly has the place of honour with all of its historical inventions: a Chappe telegraph – on video – that is in competition, from one expo to another, with Morse, which the players must learn ... . The different systems of communication necessary to the exposition itself are constantly used as points of reference.

This style of using the participation of the visitors instead of placing them before a *fait accompli* is one of the keys to Eurometrics. The little train game is always popular with the young, but it is transformed by the spirit of the expo. The children are made responsible for running an immense working model of electric trains, spread (virtually) across the twelve countries. In addition, they must come to agreement among themselves to decide on a universal time since, at the beginning of the game, their watches are taken from them and they are each given instead old pendulum clocks. They must therefore produce themselves, through negotiations at video-screen conferences with the other eleven teams, a schedule that is precise enough to regulate the movement of their models without incident.

The technical prowess required to hold the exposition in twelve different places at the same time allows a whole series of games that test the limits of these capacities and force all of the European companies to give of their very best. One game that amuses everyone, for example, is the 'bush telegraph' game on a European scale: the idea is to send a message and to translate it into the different languages, then to compare what the message has become when it is translated back into the first language. The game puts both human translators and the different translating machines into competition. It is left to the big telecommunications and computer companies to try to out-do each other. There is the same rivalry in the harmony game, where a music conductor must lead in perfect time and in the same tone, twelve voices and musical instruments situated in the twelve countries.

It is easy to see, then, that Eurometrics is not only a technical display meant to impress the visitors; rather it is the visitors who, with their requirements and demands, impress the technicians, the engineers and the manufacturers and test their capabilities. Eurometrics measures exactly the technological state of Europe in '92. Of course, all of the breakdowns and difficulties are made known, and the video and computer control rooms are all open to the public.

A large part of the expo is, in fact, devoted to technical harmonization in the area of telecommunications and computers. In a hall called 'Connection Hell', a video studio and a computer publishing station must be pieced together from a chaos of cables and plugs from different countries, and made by different manufacturers according to different standards. The most spectacular is the space game, where the children must make twelve versions of a rocket at the twelve expos during a single afternoon. Using the Computer Assisted Conception screens, the children must decide on the sharing out of the work, the shape of the rocket and the margins of tolerance. At the end of the afternoon, the rocket has to take off – on screen – and, depending on the success of the coordination, will either fly or explode. In another spot, it is the architecture apprentices who must build a house in another country of the expo, but using the standards of their own country incorporated into the computer-aided design program. There again, at each step norms must be negotiated and communication standards defined.

The role of European sponsors at Eurometrics is important. But instead of simply vaunting their products, they are asked to place at the disposition of the public the practical means by which they achieve the coordination of their productions. One manufacturer of toys has an astonishing computer-aided design program for standardizing their products; at the expo, this program becomes a game where the children can invent new characters that are compatible with the others of the company. Another manufacturer, of cars, has an integrated computer and accounting network. The system is there at the expo, working in real time and making orders, calculations or car designs. The goal is not to amaze but to show the cost and the interest of European coordination.

This same wish to educate and to communicate the difficulties is evident in all of the games dealing with standards, the 'snakes and ladders' games in particular. The players of one country try to sell their products in another country, but the potential buyers, with the help of their national specialists, must find ways to multiply the technical barriers. Since the players have direct access to European institutions and to data banks, the game intensifies quickly. There is even a replica of the European Court of Justice which arbitrates any disagreements. The aim of the expo is present everywhere: to show the importance and the costs of normalization and harmonization. For example in the game of the high-speed train which must be run from Lisbon to Copenhagen, the rails, the standards, the signalling systems and the habits of several railway companies must all be redesigned. It is the same for the 'D2MACPackage' which simulates, at reduced speed, the path of high-definition television and of their standards in Europe. At each step, the player realizes that the stock of televisions must be modified, and an industry recreated, in order to succeed in international competition.

One of the most popular games is that of comparative advantages. The players are asked to find the countries that are the most advantageous because harmonization has not yet taken place there – a fiscal paradise, for example, where workers are the least protected, or where medicine is the cheapest – but then, throughout the game, the countries react and obtain the harmonizations that destroy the unfair advantages. That is the whole cultural and educational meaning of Eurometrics: things are not over-simplified here. Europe is not a propaganda target but a large structure in the making, with all of its contradictions.

At the same time, the historical dimension is present everywhere. There is a stand dealing with the metric system, another that deals with the affair over the Greenwich meridian, and one on the *Zollverein*. Through its contributions, each country reminds the others of their own unification and the immense diversity of their efforts at unification. What is always the most interesting for the public is the contrast drawn between regional diversity, national unification and the complex yet harmonized whole that is Europe. Since the expo has chosen to show only the work and the method behind all harmonization, it never gets into the political problems – should Europe be more unified or not? – but it does, nevertheless, make the political risks more perceptible and more sensitive for many. Such is the case, for example, in the game of the ECU and that of the Great Snake. It can also be seen in the hall named 'Weighing Apples, Weighing Europe', where all of the instruments used to 'weigh' the strategic and economic strength of Europe against all of the difficulties and uncertainties of the very measuring instruments that allow us to speak of a Europe.

It is particularly interesting to see the parallel at Eurometrics between the scientific problems of the replication of experiments (the philosophy of measurement) and the most practical of problems. A series of rooms equipped for experimentation permits the visitor to ask the Twelve 'laboratories' to compare their results by replicating certain famous experiments in the history of the sciences. The disparity of the results and the difficulty in finding the causes of the errors is interesting. But it is even more interesting for the same visitor to find themselves at the other end of the exposition in a diabolical supermarket where every product is different from the others (by the price, the shape, the content, the container, the packaging, the safety standards, etc.). The players must, nevertheless, learn to manage this nightmarish supermarket by inventing their own standards of production, their own norms and regulations. This link between scientific problems and economic problems also shows up in history, which speaks both of the history of the sciences and of the history of markets and companies.

The Eurometrics philosophy can always be found in the presence of the real networks of European firms and institutions. For example, the information stands dealing with the equivalence of diplomas are continuously taken by storm. But rather than simply receiving information, the visitors attend, by way of the videophone, all of the conversations in the equivalence offices to decide on, or to find, the right equivalence. Also, all of the official meetings scheduled for the same time of year as the expo take place at the heart of the expo and are as open to the public as possible.

Well now, readers, Europeans, lovers of technological culture and of the innumerable strands that weave together sciences and societies, and who celebrated the first of

January 1993 without so much as a nod to the new European standards: were we wrong to believe in Europe? Wrong to believe that the job of standardization opens a royal path to the understanding of history and the sciences? Wrong to believe that the general public must be able to grasp for themselves both the stakes involved in this task of construction that distributes the differences and the similarities, and just how important it is?

Authors

Bruno Latour is a professor at the Ecole des Mines in Paris, and works at the Centre de Sociologie de l'Innovation, 62 bvrd Saint-Michel, F-75272 Paris, France. Mickès Coutouzis is director of the Eurometrics Project at the Ministry of Research and Technology, Almeida 1, Koukaki, Athens, Greece.

Democracy and the programme of science information in Sweden

Jan Nolin

Introduction

The provision of science information has been compulsory for Swedish universities since 1977. The requirement is termed the 'third assignment'.[1] The most widely cited argument for this is what has been called the 'democracy argument', the 'civic argument'[2] or the 'political argument'.[3] It is derived from the general idea that science is a dynamic force in society, causing social and economic change. The people of a technologically advanced society must be kept informed about important scientific findings in order to be able to function in the democratic process. In this paper I discuss several problems with an implementation of compulsory science information in connection with the democracy argument. These complications have surfaced in my study of two information units at two different universities in Gothenburg: the Technical Faculty at Chalmers, and the University of Gothenburg. I also draw some conclusions on what is and what isn't a viable base for further action if we are primarily interested in using science information to solve some crucial problems facing the democratic system in a modern advanced society.

The Swedish legislation and the local reaction

It is important to understand the vagueness of the legislative text that directs this provision of science information. It states:[4]

> The activities of the university should include the task of dissemination of research and development. Knowledge should also be disseminated about which experiences and which insights have been acquired and how these shall be applied.

This key formulation is extremely vague. It does not mention (a) *who* is required to communicate – the scientists, their departments or the university bureaucrats, or (b) with whom they should communicate – to laypeople who regularly utilize research, or to groups from the larger public.

As it turned out, this legislative move was not enforced by any kind of steering; instead it was up to every individual university to decide how they would interpret the directive. The local attempts were complemented by a federal programme organized by the Swedish Council for Planning and Coordination of Research (FRN).[5] This unit did not attempt to steer the local activity; instead they funded worthy local attempts. They also initiated a whole range of projects of their own. Most noteworthy were the *Fount* series, 40-page booklets which were very cheap and were sold everywhere, and generally in numbers of 5000–20,000 copies.

Within the universities the task was delegated to whichever unit within the bureaucracy dealt with information. The employees were given this responsibility in addition to the several others they might already have had, such as promotion, organizing parties, press contacts, editing the local university paper, the university mailing list and so on. As a result, the new and the old tasks were quickly interwoven and it became difficult to see which activity was science information. Usually a project had several purposes, some of which could have a bearing on science information.

The task of putting pressure on scientists to increase their external communication was also delegated to these science information people. But there was no possibility of forcing researchers into action. They could only try to coax, persuade and stimulate the scientists.

Two different strategies

The interpretative flexibility of the legislative text was reflected in the strategies of different universities. The two examples I studied are clear evidence of this.

For the technical school at Chalmers, the requirement meant on the one hand a strengthening of its traditional communication network, to people who applied technical knowledge in commercial enterprises. On the other hand, it also meant helping the departments with their communication needs. This meant that they were working more or less like a public relations firm. Indeed, in this respect they competed with public relation firms in winning assignments from different departments, their advantage being their closeness to and knowledge of the ways of scientists and of the Chalmers image. Departments paid for these services, so the information unit had an extra income. This enabled the unit to expand to a size that met the extra needs.

For the University of Gothenburg, the interpretation was different. This university included a large number of heterogeneous departments from humanities to medicine and natural sciences. Their communication needs were neither as fixed nor as streamlined as those at Chalmers. Also, the economic situation was not as good as at Chalmers. The information unit was smaller and not supposed to charge a fee for its services – indeed they even contributed money to scientists who were involved in interesting communication projects. Communication to people who applied research and to the general public were evenly balanced. An information professional was placed in every faculty to enable close contact with scientists. Year-books were edited which contained popularization of current research. The strategy was to try to stimulate the scientists to popularize more.

A notable tendency for the unit at the University of Gothenburg was to try to educate the scientists in communication. For Chalmers it was more a case of building up a unit which had the competence to handle the communication that scientists were unable or too busy to do. The people at Chalmers tried to develop an image like that of a successful company. The aim of the University of Gothenburg on the other hand was to project an image of a place filled with wise, serious and hard-working scientists:

a place of learning.

Both universities maintained a local focus. Both units were tied closely to the Dean's office, and at times served as his tool. Local problems perceived from the top therefore provided important input for the policy of the unit. A vital part of the activities of both units was therefore the university magazine, which served as a way of holding the university together. Other tasks were to uphold the image of the university and to work with promotional matters.

The local focus meant that the universities were not following some grand national plan. On the whole it was the immediate problems on the local level that steered their actions. This has consequences for the democracy argument, to which I will now return.

What is the democracy argument?

The democracy argument can be said to consist of two separate aims. One is to narrow the knowledge gaps in society. The other is to promote public participation in and produce public support for the decisions made on the basis of expert knowledge. A realistic strategy would seem to be to attempt to make the public acquainted with the theories with the greatest social impact, as well as with the experts that promote them.[6] It would also seem reasonable that there should be some kind of dialogue going on about policy matters that concern these theories.

An on-going discussion between society at large and the community of scientists and experts presupposes a two-way communication. The public is not only spoken *to*, but spoken *with*, and they speak back. In this way scientists receive feedback and are forced to adjust theories to new contexts. They also have to articulate their perspectives in languages other than the scientific. Members of the general public can, in turn, become participants in the policy discussion. In the Swedish concept an important idea is that not only do the people need scientific knowledge, but feedback from the general public is also essential if the intellectual climate is to be renewed and theories stretched by reality.[7]

The non-scientists of society have a right to form their own knowledge. Science and science information should not take this right away from them. Ordinary citizens should not think it impossible for them to discover something new about reality. However, it often seems that society has delegated the joy of discovering and collecting knowledge exclusively to the scientists. This perspective has been further developed by many science critics, who link the domination of nature with the domination of human beings.[8] The process of knowing and of self-reflection is handed over to experts who tell us what to think and do, and science can therefore be seen as an instrument for social control.

The scientists need feedback from society to produce better knowledge. At the same time, they too have a right to form their own knowledge. Feedback from society and the state is clearly needed, but it must not be allowed to influence science to the extent that the only areas that are investigated are those which are popular or which are of particularly obvious relevance. Still: the knowledge gap must not become too great. Today's high-tech society is steered to a great extent by the experts. There is a risk that this elite will put limits on what is knowable in society, and that no alternative advanced knowledge will be available.

In the following I discuss some problems that appear to hinder our attainment of the democratic ideal, at least in Sweden.

Impediments to democracy

1. Steering

In the Swedish concept there is no steering worth mentioning from the state. This means that local issues, at least in the cases I studied, will gain prominence. But it also means that larger policy issues will be down-played. This is tricky. It is undoubtedly good for democracy to promote local issues: contacts with non-scientists are easier and the learning potential is better, since the issues are relevant to that specific context. Taking care of local needs at the local level is a very good method of enhancing the democratic system – participation is improved. But at the same time the big issues that concern the nation as a whole are overlooked, and that is bad for democracy. Somehow, we should be able to deal with both.

Federal control might be necessary to counter the dominance of local issues. But at the same time, state steering as such is anti-democratic, since the few at the top, far removed from everyday life, tell the many at the bottom what is important.

2. Utilization

One local need that is very strongly motivated is to disseminate scientific knowledge in order to increase or make more effective its application. In my case study this turned out to be the dominating theme with Chalmers. Of course, this is a very important part of science information, but it has little to do with the democratic theme. There is a risk that this kind of dissemination will grow and become even more dominant.

3. Centre–periphery

Science information is usually built on the principle that the few people who know will inform the many who do not. There is something democratically unsatisfactory about this principle. Science information may easily reinforce the scientistic idea that the only viable knowledge is that which is produced by scientists. No knowledge produced outside this elite body can therefore compete. Scientific knowledge is produced and active within a certain context and is to some extent alien matter to the unscientific mind. 'Science is reluctant to use any language other than its own, and it remains essentially the province of scientists', wrote Diane Saunier.[9] To be able to think in a scientific way means to embark on a training process which leads to a way of thinking that is not otherwise common. Advanced scientific knowledge can often be accepted by unschooled minds, without having been assimilated. The scientist deals with theoretical knowledge, while ordinary people need practical knowledge. It is two different worlds, and transferring from one arena to the other is not unproblematic.

In addition, because distance lends enchantment, problems often look completely different when viewed up close.[10] Since scientific knowledge is legitimated by its central position and projected with such force and over such a distance, it is lent an aura of objectivity that doesn't show when viewed up close.

To counter this scientistic tendency, information must also be distributed about the relative nature of knowledge and about the social process of producing scientific knowledge.[11] But on the other hand, over-emphasize this and science will lose credibility and people will no longer listen when it really matters.

In my case studies several different kinds of attempts were made to bring children face to face with scientists in order to counter the stereotypical images of what science and scientists are. The projects were often organized on a very large scale, covering children of all ages in a certain region. The response was very good on both sides.

4. Quantity versus quality

A cornerstone of democracy is that the will of the many shall decide. At the same time it seems rational for an advanced society to take not the most popular route, but the best. And the most expert opinions are not always the most popular. A modern, capitalist society is therefore in certain ways incompatible with democracy. Long and drawn-out discussions between an elite and the masses also hinder fast decision-making. For convenience the decision-makers have to trust their experts and then inform the public, rather than putting specialized knowledge under discussion. There is also the possibility that the decision-makers trust their ideology rather than their experts. On the most controversial political issues we will usually find a scientific controversy.[12] Politicians can therefore select the theory that suits their purpose – and then inform the public.

A common thrust in my cases was the concentration on select audiences, since specialized issues tended to interest certain target groups and them alone. Complex messages could therefore be communicated, which would be more difficult if the aim were to reach a large number of individuals – the content would then have to be simplified further.

5. Idealization

One of the key problems of science information is that science is generally seen as something completely different from what it actually is. One aspect of this is that scientific knowledge can be said to be more heterogeneous, fallible, specialized and arbitrarily produced then is commonly perceived. Science can seldom live up to the enormous trust that is invested in it by society at large. In reality public support of science means that it is an idealized image of science that is supported, not the thing itself.[13]

Attempts at broad popularization are often take on a traditional dramaturgical form. This usually leads to the idealization of the scientist and the mystification of science as something obscure.[14] The success of the science writer depends on producing coverage of which scientists approve.[15] This too leads to idealization. Another problem is that absolute trust in the truthfulness of science can be a way of surviving when an individual is personally dependent on science being able to keep its every promise.[16]

In my cases the information professionals saw it as their job to communicate a good image of their university and their scientists. Idealization was therefore to a certain extent part of the job.

Conclusion

If organized science information is put into effect, then information will in different ways be channelled to various groups in society. According to the much discussed knowledge gap hypothesis, such distribution projects usually end up by strengthening the groups that are already well informed; in this way we widen the gap.[17] In Jon D. Miller's terminology, I refer to decision-makers, policy leaders and the attentive public.[18] The knowledge gap between these groups and the less privileged groups – the so-called 'interested' and the 'non-attentive' public – may easily widen for every valiant and idealistic effort that is made to spread knowledge. This is troubling, since the knowledge gap is one of the major democratic problems. We may strengthen the top, but instead we will get a wider divide between the many who know little or nothing about science and the few who are well informed. The centre can not communicate with the periphery, only with the centre of the periphery.[19]

As to Sweden, the legislative move has led to a strengthening of qualitative efforts aimed at selected audiences. But to attain the democracy ideal we need to retain the quality while maintaining the quantity. I mentioned two goals for democracy in this particular context. Science information seems very successful at the first, increasing public participation. But this must not overshadow the fact that it may very well counteract the second, and widen knowledge gaps within society.

Acknowledgement

This research has in part been sponsored by the Swedish Council for Planning and coordination of Research (FRN).

References

1   When I use the term 'science information' I refer to information about research that stems from scientists and research centres. Thus it signifies distinctly that dissemination which is controlled by the legislation.
2   Shortland, M., 1988, Advocating science: literacy and public understanding. *Impact of Science on Society*, **38**, 309; Durant, J.R., 1990, Copernicus and Conan Doyle: or why should we care about the public understanding of science? *Science and Public Affairs*, **5**, 12.
3   Hazen, R.M., and Trefil, J., 1991, Educators must accept the difference between 'doing' and 'using' science. *The Scientist*, **5**(6), 13.
4   Svensk författningssamling 1977:218.
5   Dyring, A., 1988, Public dialogue on science in Sweden. *Impact of Science on Society*, **38**, 327–336.
6   Durant, J.R., 1990, Copernicus and Conan Doyle: or why should we care about the public understanding of science? *Science and Public Affairs*, **5**, 13.
7   See *För bättre vetande: Fyra års försök med forskningsinformation. FRN-Report* 83:11, pp.127–128.
8   See Habermas, J., 1971, Technology and science as 'ideology'. *Toward a Rational Society: Student Protest, Science and Politics* (London: Heinemann); and Aronowitz, S., 1988, *Science as Power: Discourse and Ideology in Modern Society* (Minneapolis: University of Minnesota Press).
9   Saunier, D., 1988, Museology and scientific culture. *Impact of Science on Society*, **38**, 338.

10   Collins, H.M., 1985, *Changing Order: Replication and Induction in Scientific Practice* (London: Sage), pp.142–145.

11   In short, science-in-the-making: see Shapin, S., 1992, Why the public ought to understand science-in-the-making. *Public Understanding of Science*, **1**, 27–30.

12   This is a point which Trachtman sees as a basic drawback for the democracy argument. See Trachtman, L.A., 1971, The public understanding of science effort: a critique. *Science, Technology & Human Values*.

13   Brian Wynne has studied this issue: see Wynne, B., 1990, Knowledges in context. Paper presented at the conference Policies and Publics for Science and Technology, Science Museum, London, 5 April, pp.2–3.

14   See for instance Hornig, S., 1990, Television's *NOVA* and the construction of scientific truth. *Critical Studies in Mass Communication*, **7**, 11–23.

15   Dornan, C., 1990, Some problems in conceptualizing the issue of 'science and the media'. *Critical Studies in Mass Communication*, **7**, 48–71.

16   Wynne, B., 1992, Public understanding of science research: new horizons or hall of mirrors? *Public Understanding of Science*, **1**, 37–43.

17   Originally the knowledge gap hypothesis was stated in the following way: 'As the infusion of mass media information into a social system increases, segments of the population with higher socioeconomic status tend to acquire this information at a faster rate than the lower status segments, so that the gap in knowledge between these segments tends to increase rather than decrease.' Tichenor, P.J., Donohue, G.A., and Olien, C.N., Mass media flow and differential growth in knowledge. *Public Opinion Quarterly*, **34**, 159–160. More than fifty studies have attempted to analyse the process and certain revisions have been attempted, most notably in Ettema, J. F., and Kline, F.G., 1977, Deficits, differences and ceilings, contingent conditions for understanding the knowledge gap. *Communication Research*, **4**, 179–201; and Horstmann, R., 1991, Knowledge gaps revisited: secondary analyses from Germany. *European Journal of Communication*, **6**, 77–93.

18   Miller, J.D., 1986, Reaching the attentive and interested publics for science. *Scientists and Journalists: Reporting Science as News,* edited by Sharon M. Friedman, Sharon Dunwoody and Carol L. Rogers (New York: Free Press).

19   As in the analysis of political centre–periphery. See Galtung, J., 1971, A structural theory of Imperialism. *Journal of Peace Research*, **8**, 81–117.

Author

Jan Nolin is working on a PhD thesis which compares the science and the media discourse on ozone depletion. He has also published several articles on scientific controversies, nuclear waste management and social constructivism. He is at the Department of Theory of Science, University of Göteborg, 412 98 Göteborg, Sweden.

Clone, hybrid or mutant? The evolution of European science museums

Melanie Quin

The word 'museum' entered the English language in 1656. John Tradescant's *Musaeum Tradescantianum: or, A Collection of Rarities, Preserved at South Lambeth near London* catalogues the extraordinary collection of his father, the naturalist and diplomat John Tradescant the Elder who brought lilac and acacia to British gardens and opened the first public museum in the country. The Tradescant collection came into the hands of Elias Ashmole, whose Ashmolean Museum was opened in Oxford in 1683, on the upper floor of a new science 'elaboratory', and before long the collecting of curios became a social fad.

From celebration to education

The first museums were eclectic cabinets of curiosities. But those that specialize in science and technology are essentially late-nineteenth and twentieth century institutions, following on the heels of a succession of celebratory world's fairs.

The origins of the Science Museum in London can be traced to the Great Exhibition of 1851. This exhibition led to the setting up of the South Kensington Museum, which later split into the Victoria and Albert Museum, dealing with the decorative arts, and the Science Museum, which opened in its own building across the road in 1928. Working models quickly became a characteristic of the Science Museum, and in 1933 the Children's Gallery was opened, consisting 'almost entirely of demonstrations, operated by the visitor, of scientific principles'.[1] The response from the museum-going public was overwhelming: from 600,000 visitors in 1926, numbers grew to one million in 1929, and 1.5 million in 1933, making it the top national museum.

The Deutsches Museum in Munich, which was founded in 1903, set out to demonstrate the interactions of science, technology and industry during their historical development: alongside the static historical collection, machines moved and visitors activated mechanical models. The Palais de la Découverte in Paris was opened in 1937 for the International Exhibition of Art and Technology, and since then has been funded by the Ministry of Education. It immediately set out to address a broad audience. Combined with the permanent exhibits were spectacular experiments, presented by young scientists who also explained principles and answered visitors' questions. The tradition continues with over 50 live shows every day.

These three museums' populist educational mission and lively techniques influenced the policies of existing museums and inspired the establishment of others across the Atlantic (notably in Washington DC, Chicago and Philadelphia), as Sheila Grinell records:[2]

In the late 1960s, after the decade of reform in science education that followed Sputnik's launch in 1957, several institutions opened that further elaborated on the concept of interactivity. The Exploratorium in San Francisco, and the Ontario Science Centre near Toronto eschewed historical and industrial collections in favour of apparatus and programmes designed to communicate basic science in terms readily accessible to visitors. These institutions postulated that displays and programmes carefully designed to provide first-hand experience with phenomena could captivate ordinary people and, in the best of circumstances, stimulate original thinking about science.

Since the 1960s, the educational philosophy and methods of the Exploratorium, the Ontario Science Centre, and half a dozen other pioneer North American science centres have, in turn, provided inspiration for institutions around the world: new science centres have been developed, and established science museums have borrowed the techniques of the science centres to develop their museology. The science centres and new-look museums share a mission: to promote the public understanding of science and technology through exhibitions that invite active participation – 'hands-on', 'minds-on'

Medium and message

The construction of a museum is a declaration that the government and others want to influence public attitudes to science and technology, and to increase the standing of these subjects (and of scientists, engineers, etc.). The museum is a prestigious monument to that aim. The French government recently made such a declaration on a massive scale, affirming science as part of the national culture: la Cité des Sciences et de l'Industrie in Paris was d'Estaing's answer to the Pompidou Centre for the arts. Its mission is to promote 'la culture scientifique et technique' and as such la Cité includes an international conference centre and a huge multimedia resource centre and library, as well as offering 60,000 square metres of permanent and temporary exhibitions.

As Alan Morton of the London Science Museum has remarked, it is significant that at the time when museums like the Deutsches Museum, the Palais de la Découverte and the Chicago Museum of Science and Industry were first set up, the other sources of information about science and technology were, for most people, printed encyclopedias, newspapers and books – all of which are two dimensional.[3] The museum medium had the advantage of another dimension, and with it came the romance of seeing the 'real' thing on display.

Television, video, and computer games – new media of popular instruction and entertainment – now provide competition for museums. On the screen, viewers have access to images of distant people and places in a way that a museum can never match. Television can also record complex and lengthy processes, on site. Once, museums provided access to the parts of a differentiated society and were modest microcosms of industry. Today, television plays that role and conveys vivid impressions of 'being there'.

Science centres offer the opportunity to experience 'real' phenomena and experiment with 'real' processes. Unlike museums with historical collections, they have few 'real' artefacts on show. These centres face direct competition from theme parks, such as

Disney's EPCOT Center. With no mission to educate, the theme parks focus purely on fun. Theme parks make everything easy for their visitors. Thrills are experienced in physical and mental comfort; few decisions have to me made, and there are few reasons to question what one knows or believes. Science centres challenge their visitors, fully expecting them to bear the intellectual discomfort that precedes the 'Heureka!' of discovery.

Vigorous hybrids or dominant recessives?

As science centres proliferate in Europe, catalysed by the example of successful North American institutions, how will they meet their critics' challenges? Can they evolve or must they atrophy?

Traditional museums are bound to collect and conserve, as well as display. The new institutions, without collections, might be accused of being mere expressions of fashion. In the sense that every new science centre is to some extent a clone of those that opened before, the spread of the movement is an enormously successful example of niche colonization. Yet it is not self-evident that individual science centres, or the movement as a whole, will evolve further.

On a wet Saturday afternoon, families want an enjoyable social outing. For visitors who do not arrive in school groups, the science centre is an alternative to the cinema, or to going shopping, or to watching a football game. If science museums are to fulfil their educational goal, they must first compete successfully in the leisure market, and for this they need a clear identity: attractive, dynamic and distinct both from the passive medium of television, and from the escapist fairy-tales of theme parks.

Having reached the watershed point of this article, I must confess to my (subjective) impression of institutions in a state of limbo: debate within the international science-centre community reveals self-doubt and the questioning of foundations. In seeking to define a framework for the science centre of the twenty-first century, this article thus reflects the 'state of play' in Europe.

Searching for the next generation

There is a perceived need to offer, in an informal setting, scientific experiences that the formal education system is unable to provide. And most science centres attempt to provide more than an eclectic assembly of interactive exhibits. They recognize the importance of establishing links, and providing conceptual frameworks. It is popularly believed that this can be done through well-designed thematic exhibitions, accompanied by carefully crafted text labels. Part of the trick, we are told, lies in the grouping of the exhibits themselves, as Ilan Chabay, president of the New Curiosity Shop, California, explains:[4]

> The wonderful experience of being a 'barefoot empiricist wallowing in the facts' of science is not sufficient to develop a conceptual framework in which to arrange and use those facts and predict the behaviour of systems. ... One concern therefore is to develop sets of closely related exhibits in a single theme. In the museum setting, the physical layout and proximity of the exhibits can

further reinforce a sense of the relationship between elements in each exhibit and the conceptual unity among them.

Another school of thought promotes scientific problem-solving as the route to conceptual frameworks. DrewAnn Wake and James Bradburne argue that it is the responsibility of science centres to develop exhibits that allow the public to exercise their 'creative' skills of imagination, extrapolation and sensitivity to scientific process. They admit that:[5]

> We cannot pretend that these skills are entertainment, that they can be captured without intellectual effort. We do not expect people to learn them unconsciously ... still, while our exhibits may not be entertainment, they can be entertaining. They can communicate the great joy that scientific thought brings.

The 'real science' approach can be witnessed and enjoyed at Science North, in Sudbury, Ontario, which has received international recognition for its ground-breaking approach to museum teaching. Exhibits have a minimum of fancy cabinetry, and labels are largely replaced by blue-coated scientists who engage visitors in conversation and encourage them to be scientists for a few hours. It's worth remembering, though, that different people have different learning styles. Not everyone is prepared to roll up their sleeves and participate in the scientists' workshop activities. Some visitors are more turned on by spectacular displays of independent objects or phenomena.

You don't believe me? Go to your museum and watch and listen – get into visitors' heads. Or read the visitor-survey literature. This is what you'll find: as visitors move through the galleries, they selectively look at objects, interact with exhibits, and read labels. They are drawn to exhibits that are either visually compelling or intrinsically interesting to them personally. The important aspect of their activity is that it is highly selective. And each visitor's experience is different because each visitor brings their own personal agenda, and companions; each is differently affected by such physical aspects as architecture, ambience, smell, sounds, and the 'feel' of the place; and each makes different choices about what to focus on.

Interestingly, many visitors do not discriminate clearly between the time they spend viewing exhibits and their time in the shop or cafe: all these activities are part of the same event – the museum experience. Their approach is, appropriately, that of a consumer of leisure-time activities.

Exhibits or experiences?

We (and I write now as a member of the creative team of a new science centre project) must recognize that visitors are individuals, with individual needs and interests. Some will favour the deep involvement and time commitment that workshop-style activities demand; others – including the short-stop tourists – will browse and graze, pausing at exhibits that attract them. The key to the future of science museums is in meeting our visitors' varied needs and expectations.

We must also recognize the limitations of the exhibition medium. It is not a linear medium, and is not well suited to storytelling. It is particularly futile to do (badly) in three dimensions what can be done (superbly) on the pages of a magazine, or in a TV programme. Where exhibits excel is in setting a stage: we can create a stimulating environment for social interaction, active exploration and rich experiences.

Many science centres today share a family likeness, which may be traced to the pervasive influence of the 'Cookbooks' published by the Exploratorium in San Francisco. The books provide detailed exhibit recipe plans, and have given many groups the confidence and know-how to set up their own interactive exhibitions. Yet this also gives a certain 'sameness' to the exhibitions. 'See the light' at the New York Hall of Science, is almost identical to 'La lumiere démasquée' at la Villette, and both are polished versions of about 80 exhibits first developed by staff at the Exploratorium. Clones of many of those 80 exhibits can also be found in science centres large and small around the world. The 'sameness' of exhibits is, however, overlaid, on the one hand by design (monochromatic clarity in Helsinki versus pastel shades in Barcelona), and on the other by the requirements of local and/or national educational systems: the science camps, laboratory and workshop activities, teacher-training courses, evening lectures and outreach programmes are as varied as the schools and curricula that they have been developed to support.

Despite their variety of educational activities, most science centres have been designed in the style of science arcades or fairs – the Science Circuses (Carpa de la Ciencia in Spain, Discovery Dome in the UK, Questacon in Australia) are perfect examples of the type. But the fair or arcade is *not* the only model for a science centre.

Gazing into my crystal ball, I'd like to outline three metaphors for the science centre of the twenty-first century – all of them very different in physical context from the Circus. The first is the science centre as a restaurant. Following the shifting trend in London restaurants, this one is less about food, and more about a sense of occasion, of being there amid the theatre of it all – glitter, drama and smiling attentive staff. As the place fills up, the buzz mounts and everyone seems pleased to have arrived. The purists may argue (and do) that the food is simple and well cooked, not elaborate *haute* or *nouvelle* cuisine; the owners have accurately calculated that many of us like this and are comfortable with it. Most people want an event and a style, not a gastronomic revelation.

My second metaphor is the science centre as a garden. This is an artfully 'natural' succession of special spaces, in the style of the Parc de Bagatelle in Paris. It's designed for individuals to wander in, explore and make their own paths. They might find glasshouses of tropical plants, beds laid out in the Linnaean classification system, a formal flower garden with gravel paths and fountains, a grotto or some of Henry Moore's vast bronze sculptures (when these were exhibited in Kensington Gardens in 1978, visitors are invited to come up close, to touch and feel, and even to use them as a climbing frame), a rockery, shrubbery, water garden, and orchard; there may be a tree house with a long rope ladder, and a kitchen garden where you can help plant out seedlings or pick berries. The third metaphor is the science centre as an agora. This science centre fosters discussion, even formal debate. It has pavement cafes and shady colonnades, with space for informal meetings of friends and performances by buskers.

Restaurant, garden and agora are not mutually exclusive. The key is that none is a monolithic institution. There is no mega-message engraved over the entrance. Their exhibitions are designed to appeal to individuals. They prize the quality of the visitors' experience above the effective transfer of scientific knowledge and skills.

Criteria for survival versus extinction

Criteria for survival may be distilled from the three metaphors outlined above. In the style of a restaurant, the twenty-first century science centre will offer the excitement of being there. The architecturally stunning Cité des Sciences at la Villette (where it has become fashionable to record pop videos) points the way. The plans for science-theatre and art in the future Amsterdam Science Centre are similarly worth following.

In the style of a garden, the science centre will be an activity centre, at the focus of a network linking schools, universities, libraries, publishers, and related museums and zoos. Its definition of 'science' will be as catholic as mine of 'garden'. Here, Europe's rich cultural heritage is the key, and Heureka – The Finnish Science Centre – offers a twentieth century example. In Finnish, 'science' is 'tiede', derived from 'tietää' meaning 'to find the way', 'to get to know'. And because the word spans the full spectrum of human knowledge from mathematics to humanities, so also does the science centre. Visitors who came during the Summer of 1990 found a hands-on portrayal of the Iron Age, complete with a full-size replica of an Iron Age house based on archaeological information from the island of Åland. They were encouraged to join in peat-laying and making reed bundles for the roof, or digging at the nearby Jokiniemi site.

My agora is, above all, a place of social interaction, and whilst both 'restaurant' and 'garden' encourage such interaction, the 'agora' actively promotes it, in both informal and semi-formal settings. The outstanding example of this style, at present, is the Museu de la Ciència, in Barcelona. Jorge Wagensberg (scientist and museum director) has recognized that public lectures and debates, presented on the neutral ground of the museum, are an excellent medium for giving largely uninformed people an overview of the issues relevant to a scientific topic or field of research. The weekly 'Vespers in the Museum' involve the public in science as a cultural activity, encouraging active discussion and the questioning of 'real' scientists.

The crystal ball glows brightly, and I'll end on an upbeat note. ECSITE (the European Collaborative for Science, Industry and Technology Exhibitions) was established in 1990 to encourage and facilitate cooperation between science centres and museums in Europe. ECSITE is in some senses the little sister of the (American) Association of Science-Technology Centers, which is based in Washington DC. ECSITE members trade European cultural diversity and their strong sense of mission for the fruits of 20 years of American practical experience – in marketing, and wide-format films, for example. The exchange is made concrete through travelling exhibitions and staff-exchange programmes.

But at the heart of ECSITE is a forum for debate. The whys and wherefores of science centres fuel discussion in the *ECSITE Newsletter*, and at international conferences and workshops. My vision, expressed in metaphors, is a personal extrapolation from a very real search for identity – a search that is particularly active in Europe.

References

1   Day, L., 1987, A short history of the Science Museum. *Science Museum Review* (London: Science Museum), p.17.

2   Grinell, S., 1992, *A New Place for Learning Science: Starting and Running a Science Center* (Washington, DC: Association of Science-Technology Centers), pp.6–7.

3   Morton, A., 1988, Tomorrow's yesterdays: science museums and the future. *The Museum Time-Machine*, edited by R. Lumley (London, New York: Routledge), pp.128–143.

4   Chabay, I., 1989, Big exhibits from small toys grow. *Sharing Science* (London: Nuffield Foundation), pp.39–41.

5   Wake, D., and Bradburne, J., 1990, Paradox lost: re-discovering scientific creativity, paper presented at the Science Museum conference 'Policies, and Publics, for Science and Technology', 7–11 April.

Author

Melanie Quin was founding director of ECSITE – the European Collaborative for Science, Industry and Technology Exhibitions, and is editor of the *ECSITE Newsletter*. She now works at the Technologie Museum NINT, Tolstraat 129, 1074 VJ Amsterdam, the Netherlands, where she is developing exhibits for the new Netherlands Science Centre, which is scheduled to open in 1995/6 on a city-centre waterfront site.

The technological culture: opening the political and public debate

Michiel Schwarz

Technology is a way of living, a way of thinking. Today we no longer just use technology, we live it. Technology dominates our work, our food, our health, our education, our communications, the way we structure the world. That is the reality of our present-day 'technological culture'.

This idea of the 'technological culture' has been conceived in the context of a five-year project with the same title, developed at the independent Amsterdam-based 'De Balie' Centre for Culture and Politics. In shifting the social and public discourse from 'technology' and 'culture' to questions of the 'technological culture', the project has aimed to assess and open debate on the role of technological development in contemporary culture. This paper explores the 'new realities' of our 'technological culture', and describes the activities of the Balie Centre concerned with staging and promoting public and political debate on the cultural challenges posed by technology.

Technology-as-culture

The idea of the technological culture as used here differs from what – in France and elsewhere – is sometimes referred to as 'la culture scientifique et technique'. Equally, it is not concerned with accounts of scientific and technical research as cultural activities. Adopting a more anthropological concept of 'culture', the notion of the technological culture (as developed in the Amsterdam project) sets out to draw attention to technology-as-culture. It reflects the view that technology has become integral to virtually every aspect of our culture – not just in the way we think and behave, but also in the way society is organized, and the way we perceive and frame choices about our social future. Coming to terms with our technological milieu means premising our thinking on this very consciousness. In just the same way as we have come to debate politics, art or economics as elements of societal development, we need to treat technological change as a central constituent of our culture. And at the same time we should reflect on the way our norms and values impinge on the direction of technological change. The perspective of the technological culture prompts us to re-examine the way we discuss issues of 'technology and culture', and the way we make democratic choices about the technological design of our modern world.

The public and political debate on technology and culture should reflect the view that today, more than in earlier times, technical artifacts have become deeply embedded in our institutions, social relations, practices and perceptions, so much so that we have to abandon the conventional instrumental view of 'technology' that sees it merely as 'a machine'. Technologies have become 'forms of life' (to speak with Langdon

Winner) and a way of thinking (an ideology, if you wish). The notion of the technological culture underscores the all-embracing nature of our technological milieu. In modern culture, we are surrounded by TVs, videos, computers, cars, telephones, fax machines, the latest medical equipment and pharmaceutical products, and even the food we eat is full of artificial flavourings and dyed the proper colour. A web of technological artifacts structures and gives meaning to our lives. These, however, have become such a 'natural' part of our modern world that the cultural significance of technology often eludes us. Collectively we swim, one could say, in an ocean of technology. And yet, as Marshall McLuhan put it, 'the fish is the last creature capable of understanding the water'. Informing the Balie Centre's project on the technological culture is the urgency of examining this ocean, and to understand the water for what it is.

Underpinning this reflexive orientation of The Technological Culture programme is the need to redefine the relationship between 'technology' and 'culture', and to think afresh about the problems of technology as a determining force in society. And vice versa, we have to contemplate the role of social and political developments in affecting the future of technology. Put differently, we can no longer evade technology as a cultural and political issue. The Balie project, which started in 1987, consists of a programme of public lectures, discussions, performances, publications and research. It was conceived as vehicle for stimulating public debate on technology-as-culture, and to actively encourage both citizens and politicians to reassess the role of science and technology in contemporary culture.[1]

Science and technology as cultural faiths

The technological culture project was conceived in the second half of the 1980s, at a time of general optimism about technology. In the 1980s, the more critical social perspective on technology that flourished in the 1970s was replaced by a narrower, more economistic outlook. The prevailing government policies in Western European countries had come to treat technological development first and foremost as a key instrument for economic growth and industrial development. By the mid-1980s, both right wing and left wing political parties in most Western European countries were unanimous in their call for technological innovation.

In this political setting, the conception of the Balie Centre's initiatives may be seen – at least partly – as a reaction to the emergence of what some have called a 'technocratic' mode of technological culture in Western Europe. At the same time, it reflected perhaps an uneasiness with the general lack of 'alternative' approaches to the prevailing 'free market' technology policy, where technological imperatives and technological solutions were seen as the only way forward. As 'the market' and 'the economy' became the dominant contexts in which the development of science and technology was presented, social and cultural discussion on these developments were pushed to the periphery of public and political discourse.

Especially in the second half of the 1980s many Western European governments launched a 'cultural offensive' to proclaim the promises of modern science and technology. Governments and industry strived zealously to strengthen social and cultural faith in science and technology. In France this was done under the banner of 'la culture scientifique et technique'. In Britain the Royal Society has taken it upon itself to increase 'public understanding' of science and technology.

In the Netherlands, there were fewer signs of wide-scale government campaigning, but the general attitudes towards science and technology were similar. A definite tendency to boost 'cultural awareness' of technology could be observed – witness an advertizing campaign urging young people to study science as the key to a successful career ('Kies Exact!'). And since the late 1980s, the Dutch Minister responsible for 'technology policy' (Economic Affairs), has become increasingly concerned with creating cultural understanding and social support for the technological innovations that are being pushed on to the market.

This political and economic propaganda by national bodies on behalf of science and technology was ostensibly presented as a 'cultural' endeavour. Yet in the dominant picture the technology which society is expected to embrace is being presented not as a cultural phenomenon, but as an external instrument (or body of knowledge) that needs to be applied in as many sectors of the economy and society as possible. Technology is the solution, and people and culture are lagging behind, so it seemed. The official 'culture of technology' that was being promoted was a culture conducive to technology. It was hardly a serious plea for assessing the cultural agendas implied by the drive for 'technological' solutions. It did not ask what *kind* of culture we were creating in following the logic of technological imperatives.

From the perspective of those concerned with cultural and social developments – the context in which the Balie Centre for Culture and Politics operates – the dilemmas raised by the realities of our technological milieu and our technological life style clearly required a redefinition of 'technology' and 'culture'. Moreover it required a new perspective that would allow technology to be debated in social, political and cultural terms.

From technology to technological culture

Most of the recent efforts to inform 'the public' about technological developments – be they by science journalists, in museums or in television documentaries – have been committed to (and frequently reinforce) an instrumental view that treats the technological and cultural realms as separate domains. Rather than examining the nature of our technological 'milieu' – McLuhan's water – those involved in informing the public have been largely content to treat technologies as distinct entities that ought to be explained in their instrumental functions. The Balie Centre's programme took a different approach. Instead of focusing on technical developments *per se* – as traditional science journalism has tended to do – it took up the challenge of examining and debating the social and political roots of the technological culture. It has set out to lay bare the landscape of our technological milieu, and in doing so it has tried to shift the public and political discourse from 'technology' to 'technological culture'. It draws attention to the fact that technological change has increasingly become an end in itself, a value that comes to dominate our thoughts and actions. Technology has ceased to be merely a 'tool', a means to an end. It has now become both our environment and the guiding principle in modern society. 'Techno-logical' concepts seem to define our vision; they colour the way we perceive problems and define the way we devise solutions. Technological imperatives have merged with our norms and values. This view of technological culture, then, is a two-sided one. Firstly, technology is our 'milieu', and in this sense we live *in* a technological culture. And secondly, technological change has become the dominant force in shaping the values, norms and expectations that together structure our thoughts and actions. In this sense Western culture *is* a technological culture.

Developed from this perspective, the Balie programme centred around two focal points. First, it aimed to inspire new thinking on technological culture by taking a anthropological view and to confront people with important elements that make society a technological society. Among other things, it has drawn attention to issues such as the ethical dilemmas raised by medical technology and genetic engineering, the role of computers in education and learning, the institutions behind technological developments, and the transforming character of technology in art and culture. Second, it set out to examine technology as a determining factor in our culture, making it an explicit issue for public and political debate. It stages debates on questions of control in directing technological developments towards socially viable paths, thus formulating the challenges this poses for democracy and the political process.

To date, the Balie Centre has presented more than forty public lectures, meetings and debates and has published six books in the context of The Technological Culture project. Topics have covered a wide range and have included both specific technological developments (genetic engineering, biotechnology, media technology, food technology, medical technology, military technology, artificial intelligence) and broader cultural and political issues where technological developments play a significant role (education, Third World development, city planning and architecture, art and design, democratic control, science policy, advertising, nature and the environment).

The programme has employed a variety of means: lectures and talks, political discussions, informal seminars, films, dramatized readings, theatrical demonstrations, installations, visual presentations and interactive displays. In addition to the public meetings, Balie has organized a number of closed sessions with people from politics, science, industry, public interest groups and government. In many instances it has proved to provide an unique meeting place – a 'border zone' – where different 'concerned parties' could confront each other and create dialogues on social and political issues surrounding technology. Participants in The Technological Culture programmes have included academics involved with social and political research on science and technology, politicians, representatives of social groups, writers, artists, journalists, film makers, scientists, engineers, policy-makers and environmentalists.

Unlike more conventional, 'objectivist' public information programmes on science and technology, The Technological Culture project was aimed specifically at generating and inviting public debate. It represents, one could say, a 'dialectic approach', in the Greek sense of the systematic critique of assumptions, arguments and conclusions. One the main innovative aspects is its outspoken viewpoint of science-and-technology-as-culture, and its concern with the social and political agenda. Especially as compared to more conventional approaches to 'public information on science and technology', the programme has been explicit in initiating informed debate on contemporary issues of technology. It could perhaps be said that The Technological Culture programme compares to conventional approaches for enhancing public understanding on technology in the same way as the so-called 'new journalism' relates to traditional reporting.

The new cultural politics of technology

The Technological Culture programme, while not directly aimed at influencing governmental policy-making, nonetheless has implications for the way we think about technology policy. If, as we argue, we live in a technological culture, we need to go well beyond the current instrumental view of technological change. In this respect, technology policy has been rooted in an out-dated picture of technology that is – rightly – challenged by the view of technology-as-culture.

The last two decades have seen a number of distinct orientations in national technology policies. During the 1970s governments in West European countries saw science and technology largely as 'problem solvers'. In this period, the aim was to gear science and technology to societal criteria. In the 1980s, government priorities for science and technology changed markedly. Technology policy was largely reduced to industrial policy. Government measures came to focus on promoting technological innovation in industry, and on stimulating research and development in a number of key high-technology sectors such as information technology, biotechnology and new materials.

This prevailing concern with social and economic goals was underpinned by an instrumental outlook on technology policy. The broader social and cultural implications of our collective faith in technological innovation received relatively little attention. Notwithstanding the emergence, in the 1980s, of national organizations for technology assessment (set up to examine the social impacts of technologies), technical imperatives and technical fixes continued to dominate policy thinking in most European countries.

Towards the end of the 1980s, awareness grew that the social dimensions of technology had been undervalued in government policy. Yet, a significant shift in political perspective was not on the cards. Policy institutions largely ignored the deeper role of technical change in social and cultural change. New technologies are likely to affect our lives in a wide range of areas that constitute what we call 'culture' – in education, health, the community, industrial relations, food, human interaction, the city, the countryside and the arts. In this respect, politics so far has failed to come to terms with the reality of our technological culture. Politicians and policy-makers still tend to talk of 'technological policy' as if it were a separate field of governmental responsibility. Measures for technological innovation are spoken of in the same breath as departmental policy on transport or health. In a society that has become a technological society, no area of government can afford to be impervious to the social challenges of technological change. Whether education, health services, or the environment are at stake, sooner or later choices will have to be made as to the role we envisage for high-tech solutions and technological innovation.

Some shifts in policy thinking have started to take place, notably in the area of 'technology assessment'. In the Netherlands, we saw the creation of the Netherlands Organization for Technology Assessment (NOTA). In the field of 'public understanding' the Foundation for Public Information on Science, Technology and the Humanities (PWT) was set up. In the context of the Organization for Economic Co-operation and Development (OECD), a 1988 report on the social aspects of new technologies argued the need for a broad-based consensus about the impact of new technologies on the social fabric. It recommended the further development of different forms of technology assessment, with the basic aim 'to provide information to those concerned, to promote and participate in constructive public debate in a wide circle of institutions'. It rightly established an explicit connection between technology policy, technology assessment and constructive public debate. In practice, however, few of these principles were adopted by governmental bodies, or by social organizations or political parties for that matter.

It is in this context that the Balie programme could make a constructive contribution. By inviting debate on the current concern with new technologies and by critically discussing the bias towards 'technical fix' approaches, The Technological Culture project inevitably raised questions of politics and political culture. To what extent do we choose to make technological developments a guiding principle for health services, education, industrial organization, and the pattern of our work and leisure activities? And can our existing political and democratic institutions deal with such questions? Such questions continue to be of real significance in contemplating the future of modern societies. But it seems that these are hardly taken up by the political institutions to which modern democracies have entrusted decisions on the design of our society and our culture. In Europe, as elsewhere, the cultural agendas implicated by a technology-based notion of social progress are largely being ignored in the political debate.

The Balie project politicizes technology in a fundamental manner. It creates new openings for political and social debate. Its aim is not, however, to polarize the issues. In the technological world, a black-and-white choice of being 'for' or 'against' technology has become a false dilemma. If you ask who's right, you're wrong. In a technological culture it is meaningless to choose between a 'defensive' and a 'no-nonsense', pragmatic policy, between controlling technology and letting it go its own sweet way. Instead we need to find new ways of channelling different concerns about the social role of technological developments, and to search for meaningful criteria to judge whether or not consensus on the desirability of certain technological developments can be reached.

We are indeed facing cultural dilemmas – dilemmas of 'technological culture': about the shape of people's future in a technological world; about human identity and community; about participation; about ideas of progress. The Balie Centre's project is an attempt to provide an impetus for new thinking, new discourses, new approaches. By creating a public platform for informed debate, the dilemmas concerned can be explored from the premise that the social and political debates on technology will have to inform a new politics of technology that is conceived as part of *cultural* politics, in the sense that it is concerned with designing the future of our culture. Hence, not technology *per se*, but the shape of our technological culture is the key item on the agenda. The challenge is to get to grips with our irrevocable commitment to our technological milieu. The subject of the debate (or is it a soliloquy?) is the design of our technological world. And such a reflexive debate is only meaningful if we question the nature of the technological culture of which we are all so much a part.

Reference

1   The 'De Balie' Centre for culture and politics is an independent, non-profit centre set up to examine and promote discussion on contemporary social, cultural and political issues. Balie is a theatre, a publishing house, a meeting place, a political laboratory and a cafe. On the basis of

research Balie presents a wide range of activities such as lectures, debates, theatre performances and literary talks. It is a gathering place for artists, dramatists, politicians, journalists, scientists, writers and others. By bringing together people with different perspectives, Balie invites interaction between the world of culture and the world of politics, aiming to stimulate new insights through controversial programmes and publications. Balie is supported financially by the Dutch Ministry of Culture, the City of Amsterdam, private sponsorship and additional income from research and advisory services, primarily from government bodies.

Author

Dr Michiel Schwarz is a political scientist and sociologist working on the cultural analysis of technology, science policy, technology assessment and social debates on technology, environment and culture. He leads 'The Technological Culture' programme at De Balie Centre for Culture and Politics, Amsterdam. He is the co-author (with M. Thompson) of *Divided We Stand – Redefining Politics, Technology and Social Choice* (London and Philadelphia, 1990) and editor of the anthology *De technologische cultuur* (The Technological Culture) (Amsterdam, 1989). He holds a PhD in science policy from Imperial College of Science and Technology, London, and is a Fellow of the Transnational Institute in Amsterdam.

Science 'worth a journey'

Charles Tanford and Jacqueline Reynolds

Many years ago when motoring through France we drove some kilometres off our set route to a spot highly recommended by a well-travelled colleague. Our ever-present *guide vert* listed Vézelay as 'worth a detour', but we were assured it was, in fact, 'worth a journey' – and indeed it was (and is). Ste-Madeleine, a grand romanesque basilica built in the twelfth century, is one of the great sites in a country more than generously endowed with architectural beauty. History is present here too, since Vézelay was the site of the preaching of the second crusade in 1146 and the place where Richard the Lionheart and Philippe Augustus pledged their own support for the liberation of Jerusalem in 1190. Every British schoolchild learns about Richard the Lionheart, and the Crusades are part of the cultural history of all Europe, but the beauty of the basilica and the climb to it through the narrow streets of the old town bring a new dimension to our historical perception, and create an indelible memory.

How many of those who stand and admire Ste-Madeleine and think back on the history of the town itself know that only a few kilometres away on the banks of the Yonne is the birthplace of Jean Baptiste Fourier? They may go to Auxerre to see the old city with its narrow, winding streets and lovely Church of St Etienne, but they probably won't pay too much attention to the plaque on the walls of the municipal museum paying tribute to one of the great physicists of Napoleonic France. And why should they? Do any of them recognize his name or have any idea what his work means to their everyday lives? Does Fourier have anything to do with French 'culture' as the word is understood today? Physics and engineering students all know him of course, but physics and engineering are technical pursuits, dull, unromantic, lacking drama, uninteresting to the visitor who likes a pretty view and a stirring historical tale to go with it. But is that true?

In fact, all the ingredients of drama are present in Fourier's life (and even its subsequent commemoration) to appeal to the public taste – national history, war, tests of personal friendship, courage in the face of the enemy. Fourier was a close friend of Napoleon, a member of the team of savants that accompanied him on the conquest of Egypt, but the course of history put the friendship to severe test a decade later – there is human interest galore. (One of the tablets on the wall commemorates the Egyptian years, Napoleon in front, leading his troops, Fourier on the sidelines, writing things down.) The Auxerrois put up a statue to commemorate their famous native son in 1849. It would still be there, but for the German occupiers in World War II, who needed metal for their guns and melted down the statue. But they did not get all of it, for the mayor in a stealthy midnight raid saved two of the statue's bas-reliefs, and they are the tablets now mounted on the walls of the municipal museum, in the Place du Maréchal Leclerc.

'Science is the cultural activity for which the twentieth century will be pre-eminently remembered,' science writer Walter Gratzer tells us; 'its Golden Age is in full flower.'[1] Most of the origins of this golden age are to be found in Europe, in nearly every part of it. Scientists travelling in Europe, whether consciously seeking roots or not, will find records of their predecessors, sometimes in museums or in homes preserved, sometimes in the form of grand monuments or other public tributes. Or the remembrance may be just a simple tombstone, perhaps with flowers on it from an unknown disciple. Auxerre is just one small example; there are hundreds of others from one end of Europe to the other.

Recently, we set out on a journey through Europe to find these scientific roots and to communicate our experiences not only to our colleagues but also to laypeople, in the form of a book, a kind of scientific travel guide.[2] Our aim was to arouse enthusiasm for this great scientific heritage. We found, as we have already stated, how international the quest for knowledge has been, so that almost every country has made some contribution. Nevertheless, each country has its own unique traditions and style, in the manner of doing research, extent of public support, royal patronage and so on. Science is ultimately the product of people, conditioned by the countries in which they were educated or subsequently chose to live and by governments that could help or interfere. Scientific ideas can appropriately be called 'international', but never the people who created them nor the circumstances in which they were created. The national ethos continues to this day – the celebration of past achievements is muted in some countries, flamboyant in others. All of this adds zest to the journey.

Our scientific traveller must of course come prepared or even eager to learn, and it needs to be said right from the start that there is much to be learned. Science has not yet been integrated into our culture in quite the same way as art or architecture or literature or music; even scientists themselves are more often than not abysmally ignorant of the history of their subject. Let us cite, for example, Sidney Brenner, former director of Cambridge's Laboratory of Molecular Biology, who wrote as follows in 1985:[3]

> For most young molecular biologists, the history of the subject is divided into two epochs: the last two years and everything else before that. The present and very recent past are perceived in sharp detail but the rest is swathed in a legendary mist where Crick, Watson, Mendel, Darwin – perhaps even Aristotle – coexist as uneasy contemporaries.

It may read like an exaggeration, but it is really quite true, not only for molecular biology, but for all highly active fields of science. The new student or researcher (and, through the popular press, even the onlooker) is confronted by current problems, with a minimum of background – barely enough to understand the jargon of the particular field – and denied the stimulation that a historical perspective could provide.

It was not always so. Science was not always apart from other cultural activities: our forebears included science as an essential interest of any educated individual and did not hesitate to bring it into their drawing rooms. Monarchs took an active interest in it and writers and philosophers argued about it. In Britain, for example, there was Thomas Carlyle, opinionated writer, famous (or infamous) for his convoluted style, a man at the core of London's intellectual community in his time. He lived and entertained at a house in Chelsea (still there and now a museum in his honour), and one fascinating point about his guests is the mixture of literati and scientists. Dinner guests included everyone who was anyone. Charles Darwin and Charles Dickens might well sit at the same table. Charles Lyell was a frequent guest, as was Charles Babbage, retrospective 'father' of computers. There was much verbal sparring and acrimonious argument and Darwin, in particular, had little use for Carlyle: 'I never met a man with a mind so ill adapted for scientific research,' he wrote in his autobiography. Yet this mix of dinner guests was surely an advantage for original thought – what could be more dull than a company of people who all think alike?

In France, seventeenth century monarchs established learned academies, the famous Jardin du Roi (now Jardin des Plantes), and the Paris Observatory (still there to visit) where the Dane, Ole Rømer, first measured the speed of light – all expenses paid by Louis XIV. We can imagine what a revelation this work was to a world that had previously believed the transmission of light to be instantaneous, with no lapse of time between the departure of a signal from one point and its arrival somewhere else. When Rømer returned to his native land, King Christian built him a new observatory there, and Rømer later became mayor of the city of Copenhagen and subsequently chief of police.

Similarly, the French *philosophes* of the seventeenth century were embroiled in scientific debate. Voltaire expounded on the wonders of Newtonian mechanics, and his mistress, the Marquise du Chatelet, translated Newton's *Principia* into French, an undertaking that occupied her for more than a decade. Go to Lunéville and see how the town honours Voltaire and the Marquise, and visit the tomb of this remarkable woman. And Napoleon – already mentioned in connection with Fourier – was a great patron of all science. He was fascinated by Alessandro Volta from Italy, for example, and his demonstration of the marvellous 'piles' by which a continuous source of electricity could be generated. Napoleon gave Volta a gold medal and other honours.

So why today the gap, the 'two cultures'? Why are the popular accolades reserved for tenors and sopranos, playwrights and actors? Are the media to blame? They try to educate us about the wonders of current science, but they stress the novelty, the latest breakthrough in curing disease, in manipulating genetic material, in assessing the evidence for and against the big bang theory of the creation of the universe. They emphasize the mystery, the genius of the scientist, the illusion (at least we think it is an illusion) that it's all happening *now*, without precedent, as a gateway to a glorious twenty-first century. In fact, there is continuity with the past, and progress is no faster today than it ever was. And this continuity is the real key to public understanding: today's research on AIDS mirrors Pasteur and Koch's researches a century ago; today's geniuses stand on the shoulders of their predecessors.

A journey such as ours can be a cure for this disease of historical short-sightedness and a stimulus for learning about cultural areas of which we have only vaguely heard. (Science is no different in this respect from, shall we say, the history of the holy crusades of which Vézelay reminds us.) There are hundreds of places in Europe that one can visit where one can learn about our scientific heritage. Most of them require no special expertise, and are geared to laypeople as much as to professionals. They range from ancient history to modern, from the memorial to Aristotle's birthplace on remote and rugged hills in north-eastern Greece, prominently signposted on the main road from Thessaloniki to Mount Athos, to the place in Cambridge where Watson and Crick discovered the DNA double helix in 1953 and thereby ushered in the modern era of molecular biology.

There are birthplaces that tell a story. Isaac Newton's was in a farmhouse near Grantham in Lincolnshire. Louis Pasteur was the son of a humble tanner in Dôle, in the Jura, and the unpretentious house, on the edge of a canal, is now a small museum.

Pasteur came back here in 1883, when a commemorative plaque at the *maison natale* was dedicated, and made a little speech:

> Oh! my father and mother. Oh! my dear departed ones, who lived so humbly
> in this little house, it is to you that I owe all!

And it is to his parents that he ascribes the intense patriotism that was so characteristic of him. Their care was 'to teach me the greatness of France', and it was their inspiration that led him to associate the greatness of science with the greatness of his country.

And, equally revealing, we have the tombs. Isaac Newton was buried in Westminster Abbey in London, close to Shakespeare and other literary 'greats' (no 'two cultures' here). And Louis Pasteur has the grandest tomb of all, a mausoleum in the basement of his former home on the grounds of the Pasteur Institute in Paris, decorated with glorious glass mosaics that illustrate the high points of his career. How many Parisians have been to see it, and how many of the visiting scientists who come to the Pasteur Institute for research or lectures?

For a glimpse at creativity in physics, a grand place is the Einstein museum in Bern. Hundreds of visitors each day admire the famous mechanical clock in the heart of the old city, in the tower of the thirteenth century West Gate, with its chimes and numerous painted figures, emerging into view to herald the stroke of the hour. How many know that Albert Einstein's greatest papers – relativity, the photoelectric effect, the theory of Brownian motion – were all written just a few doors down the street, and all published in a single year (1905)? Do even most physics students know that Einstein was not at the time a professor at a university or (as he later would be) a director of a grand institute, but a humble underling of the patent office, lacking any illustrious colleagues or mentors? (Yes, we all know that he was at one time a patent official, but are we aware that that was the very time when he created the physics revolution?) The Einstein house is a place to appreciate that genius can be had without pomposity, without grants of money, and even with a sense of humour, for Einstein and a couple of young friends scoffed at the establishment as they talked about physics and had a good time doing it – the museum gives us photographs and quotes the jokes.

Likewise in biology, history has often been quirky. It used to be thought that simple living organisms created themselves spontaneously from decaying organic matter. Francesco Redi disproved the thesis (for the common house fly) in experiments done under the patronage of Leopoldo de Medici in Florence, and, appropriately, there is a fine science museum just a few steps from the Uffizi art gallery in Florence. How many art lovers go down the road to see it, to round out their lessons in Renaissance culture? Brno, in the present Czech Republic, is where modern genetics was born, by the hands of Gregor Mendel, a monk in an Augustinian monastery. There is a small museum there, too, and it shows us that 'monks' at the time could be quite worldly in their outlook and activities.

Sometimes it is appropriate to memorialize places – caves, rocks, and mountains – rather than people. There is Neanderthal, near Düsseldorf in Germany, where the famous skeleton was discovered – now it's a park, with hiking trails and places to drink beer while absorbing the rough jolt that the discovery gave to our thoughts about our past. Then there is a rock, the Roche Tulière, in the Auvergne, with the definitive example of vertical columnar basalt in the midst of former volcanic craters, which forced eighteenth century geologists to abandon the notion that floods (like Noah's flood in the Bible) are the all-important formative force in shaping the appearance of the Earth's surface. And on the rock walls of Bottaccione Gorge, near the Italian town of Gubbio, we can all clearly see the thin strip of soft clay in the sedimentary layers, evidence for a rain of debris from the impact of an asteroid with the Earth, 60 million years ago, which may have wiped out the dinosaurs.

There are also grand museums, of course, some of them very fine, where huge efforts are made to explain the totality of scientific facts and theories to laypeople with even the slightest curiosity and interest. But they tend by their very ambition and scope to overwhelm the visitor and can, in some cases, come through as an extension of formal schooling. In the final analysis there is no substitute for the sheer pleasure of 'being there'. There is something special about standing at the actual place where history was made, or seeing the actual tools with which it was made.

Our journey has led us to reflect on the strange place that science occupies in today's world. It is fragmented within itself – microbiologists live in a separate cosmos from physicists – which is the opposite of what was true in olden times, when all inquiry into nature fell under the single heading of 'natural philosophy'. More disturbing, however, is the external gulf, the alienation between all of science and laypeople. We use the fruits of science, and even eagerly await each year's new crop, but we shy away from the roots from which they spring. Too difficult to understand, is the common attitude; you need a PhD to know what's going on. Where is the 'humanity'?

What is the remedy? Not formal education, says Michael Shortland (in a review of *Alliage*, as it happens), calling it 'tried and failed'.[4] That seems too harsh a verdict, for formal education is surely an essential first step in the establishment of cultural understanding. What is, however, needed is that formal education be supplemented, assimilated into the mix of anecdotes, songs and stories that constitute our popular culture. We need to become as familiar with science's heroes and villains, triumphs and scandals – the essential 'humanity' – as we are in the case of political and social history and the arts. And we need to appreciate and communicate the intimate relationship between the arts and science. For example, how clever we were at the time of the crusades – while Richard the Lionheart and his allies were setting out to slaughter Saracens to protect Christianity against the infidel, other Christians, more scholarly types, were busy *learning* from Islam, absorbing its wealth of scientific knowledge, transcribing and translating it, providing the actual entrée for natural science into Western culture. One could fill pages with later examples.

Jacob Bronowski extols the virtues of television as a medium for communicating the excitement of scientific discovery: it is 'powerful and immediate to the eye, able to take the spectator bodily into the places and processes that are described, and conversational enough to make him conscious that what he witnesses are not events but the actions of people'.[5] All true, but we suggest there is another medium that can be even better – seeing for oneself, with its extra dimension of space and longer time to absorb and reflect. Travel and tourism have become a normal part of all our lives, and for many travellers exposure to foreign cultures and new ideas is an integral part of their experience, an important element of what would otherwise be empty holidays. Why not add 'science' to the list? In a decade or two, perhaps, every tourist guide will have its scientifically oriented places 'worth a journey' or 'worth a detour' – as much part of the tourist itinerary as monuments to wars and battles or to artistic creativity.

References

1   Gratzer, W., 1989, *Literary Companion to Science* (Harlow: Longman)
2   Tanford, C., and Reynolds, J., 1992, *The Scientific Traveller* (Chichester: Wiley).
3   Brenner, S., 1985, *Nature*, **317**, 209
4   Shortland, M., 1992, review of *Alliage. Public Understanding of Science*, **1**, 236.
5   Bronowski, J., 1973, *The Ascent of Man* (London: BBC)

Authors

Charles Tanford and Jacqueline Reynolds are Professors Emeriti at Duke University (USA) and now live at Tarlswood, Back Lane, Easingwold, North Yorkshire YO6 3BG, UK. Both have broad experience in chemical and biological research; both have received Guggenheim fellowships and other awards. Charles Tanford was Eastman Visiting Professor at Oxford University and Balliol College in 1977/8. They are at present writing a sequel to *The Scientific Traveller*, a travel guide to scientific Britain.

Epilogue

The difficult art of forecasting

Karel Dekk

Bureaucracy will ruin us ... . His face bore the signs of inexpressible weariness as Ponti removed his liquid crystal eyepiece and slumped back in his seat. A titter came from the seat in front. Had someone recognized him? Repressing a sigh the Secretary of State lowered his eyepiece, despite the fact that he was incapable of giving even the least attention to the stream of exchange rates being transmitted directly from the stock exchange in Potsdamerplatz. 'I'm tired,' he thought, lighting up a Moscow Special. 'As soon as this business is over, I'll take Natacha and the boys and spend ten hours at the PseudoZen Institute. Deacon won't be able to refuse me that ...'.

That morning the serving President of the Community had been holographed into Ponti's Amsterdam apartment. His face had been pale, his voice curt.

'Carlo,' he said, 'we have a problem.'

'A problem, Sir?'

'The elections are in two days' time. You know as well as I the results of the last projections.'

'If I may, Sir, this is nothing new.'

Deacon fixed him with a cold stare. 'That's what the Party has been telling me for the last six months. Well, I no longer agree with it. You and me, my dear chap, we are going to take things in hand and succeed where these fools have failed.'

'You mean to say ... turn round public opinion?'

'And how!' thundered the President. 'Now then, what have we done to alienate the conservative electorate on this issue?'

Ponti shrugged his shoulders and replied: 'The Tuttle Report?'

'*Europe of the Future*?'

'Yes, Sir. We have followed its recommendations from the beginning of the decade and established most of the anticipated infrastructures. Unfortunately ...'.

The presidential hologram was crackling with impatience. 'Yes, well?'

'Forgive my frankness, Sir, but I don't think the majority of people have been with us on this one. All these measures envisaged by Tuttle – the construction of towns under domes, the installation of atmospheric factories in urban settings, the rehousing of worker populations underground, the construction of a gigantic aqueduct stretching from the east to the west of the continent – and I'm quoting the actual words of the Report here – such measures have not been understood. Not to mention those measures we've had to postpone: the division of the European day into fifty hours composed of twenty seven minutes each, the establishment of an independent governmental committee in geostationary orbit, the bombardment of the continental substratum ...'.

'Enough! Enough!' The President was deathly pale. 'Did we really do all that?'

'I'm afraid so, Sir.'

'But why? Tuttle is one of the best experts in the Forecasting Department. He would never have submitted this Report without including recommendations for a global project ...'.

'Without doubt such a project does exist, Sir, but – and this is also one of the reasons why public opinion wanted us to do it – we were never able to lay claim to it. For the simple reason that Tuttle did not do it. This explains the disastrous technocratic image of the government and, if you'll forgive me, your personal popularity ...'.

Deacon raised his hand. 'Don't say any more, Carlo. You are my Secretary of State for Public Relations, my aide-de-camp so to speak. Find Tuttle and extract a D2-Mac statement from him. Something brief, simple and concrete. In two days' time Europe will know the name of the awe-inspiring design which will transform it.'

These words were echoing round Ponti's mind as he left the TTGV terminal in Prague. Without a glance at the airborne advertisements proclaiming the merits of a stay on the Moon, Mars or asteroids, the Secretary of State headed for Stanislas Square, and entered the Forecasting Department building. Tuttle's office was on the seventh floor. He knocked on the door, and it was opened by an old man whose pleasant face was hidden behind antique corrective spectacles.

'Mr Tuttle?'

'What?' The man leaned forward and threw an angry glance at the sign gleaming on his door. 'Damn and blast it! Will they never change this name plate?'

Then, turning to Ponti, he explained: 'A mistake which is at least fifteen years old. My name is Buttle. If you're looking for Tuttle, the expert in industrial geography, he's at the end of the hall.'

Ponti half-turned, then changed his mind. *Buttle*? Retracing his steps, he asked hesitantly: 'Pardon me. Could you please tell me your profession?'

The man smiled a very modest smile. 'Astronomy. And by way of secondary activity, I'm a specialist in the colonization of the solar system. And yourself ...?'

'I am a member of the government.'

'Carlo Ponti, isn't it?' Buttle shook his head in agitation. 'In that case perhaps you can give me some information. It's been ten years since I submitted a Report to President Deacon and ...'.

Ponti suffered a sudden attack of giddiness. *Bureaucracy* ... . Placing his hands on Buttle's shoulders, he heard himself stammer: 'A Report?'

'Yes, ordered by ESA. Oh, it wasn't a big deal. Just a few ideas for colonizing Europa ... . I beg your pardon? Oh come now, *Europa*! The fourth largest moon of Jupiter ...'.

Author

Karel Dekk is a science fiction writer. This was written for a competition organized by the European Convention of Science Fiction Writers on the subject of 'Europe of the Future', and won first prize. It was first published in *Alliage* in the autumn of 1990.